Nuclear Engineering Division 2018

Core Programming Area at the 2018 AIChE Annual Meeting

Pittsburgh, Pennsylvania, USA
28 October - 2 November 2018

ISBN: 978-1-5108-7621-7

Printed from e-media with permission by:

Curran Associates, Inc.
57 Morehouse Lane
Red Hook, NY 12571

Some format issues inherent in the e-media version may also appear in this print version.

Printed by Curran Associates, Inc. (2019)

For permission requests, please contact AIChE
at the address below.

AIChE
120 Wall Street, FL 23
New York, NY 10005-4020

Phone: (800) 242-4363
Fax: (203) 775-5177

www.aiche.org

Additional copies of this publication are available from:

Curran Associates, Inc.
57 Morehouse Lane
Red Hook, NY 12571 USA
Phone: 845-758-0400
Fax: 845-758-2633
Email: curran@proceedings.com
Web: www.proceedings.com

TABLE OF CONTENTS

Some abstracts in this book refer to images that were not made available by the publisher

540719 Reflections from a Chemical Engineer and Life's Lessons Learned Throughout a Career

Abstract:

Robert S Eby (Bob) will present a summary of his major technical accomplishments over a 45-year career in radiochemical engineering. Throughout this career, he has been involved in many major national programs and learned numerous lessons that have been invaluable not only in his work life as a chemical engineer/manager but also in his personal life. The goal of this talk is to reflect upon accomplishments and challenges of a typical chemical engineer, and also for every one present to hear at least one lesson learned that may be of value in either their professional or personal life.

MOBILE BORATION SYSTEM

Ryan T. Vanston P.E.
Westinghouse Electric Company LLC
Cranberry Township, PA, USA

Matthew M. Swartz PhD
Westinghouse Electric Company LLC
Cranberry Township, PA, USA

John Lojek P.E.
Westinghouse Electric Company LLC
Cranberry Township, PA, USA

Abstract

After the accident at Fukushima in March 2011, the nuclear industry developed strategies and regulatory requirements with the goal to enhance plant safety and defense-in-depth coping strategies at a global level. The specific regulations varied between countries and included a range of response strategies, from mobile equipment being stored in diverse, protected buildings to separate permanently-installed response systems. With this new ideology of being able to cope with extreme and unknown hazards beyond those already evaluated as part of the design basis of the respective sites, the design of ancillary nuclear response equipment and subsequently, portable response equipment, emerged.

Critical aspects of post-accident response would be to ensure core sub-criticality and continuous core cooling. In pressurized water reactors, sub-criticality relies partially on the use of borated water within the reactor coolant system. Certain postulated scenarios may require borated water sources to be protected from external hazard events as well as require the ability to generate borated coolant indefinitely. Desirable features of a system that is able to produce borated coolant are easy transportation, straightforward commissioning, able to provide a continuous borated discharge stream.

Westinghouse has designed, manufactured, and tested a mobile unit that is able to continuously generate borated coolant using an automated platform, minimal site interfaces, and efficient design characteristics. The design takes advantage of a multiple, high-concentration batch-to-dilution style system in order to produce a continuous discharge. Because, energy efficiency and timing are crucial system specifications, there are custom designed and tested heating sources as well as a programmable automation controller (PAC) that monitors and controls all of the instrumentation, feedback signals, and automated actions. The system reduces the time required to produce a specified volume compared to other mobile systems from days to hours, and can be used outside of controlled areas. The current system design uses boric acid as the neutron poison but the system can be designed to accept other powder, such as sodium pentaborate, as required by a reactor design and accident scenario.

The purpose of this paper is to outline and describe novel aspects of the Westinghouse system, describe the system's ability for enhancement of plant safety and risk mitigation, and identify potential future areas of use that would enhance plant operational efficiency. The system has been delivered to a site in the United States for a defense-in-depth application.

Introduction

Historically in the nuclear industry, boric acid solutions are batched in a non-continuous process. The mobile boration system (MBS) was initially designed to produce a continuous boric acid solution at the request of a utility based on the events at Fukushima in order to enhance defense-in-depth capability and improve overall plant safety. Subsequent industry and regulatory events led to design modifications and customization efforts prior to complete product delivery in September 2016. The novel and unique aspects of this system are discussed herein which have been captured in United States Patents (References 1 and 2).

Technical Overview

The MBS is a portable, on-demand system that supplies a target discharge source with borated coolant either during normal plant operations or as support for a plant's coping response to an extreme external event. The system design allows for transportation in a self-contained unit that can be moved intact on a semi-trailer, simplifying transportation, storage, and deployment. This unique delivery system is becoming more common throughout the industry as advantages in manufacturing, testing, delivery, and versatility are realized as outlined in Reference 3.

The system requires four main inputs: preferably high-quality water, a 480V three-phase power source, boric acid powder and an optional pH control agent powder. From these inputs, the trailer-sized system synthesizes a solution of the desired concentration at a specified flow rate from 2,000 ppm boron at 150 gpm (34.1 m^3/hr) to 7,000 ppm boron at 30 gpm (6.8 m^3/hr), depending upon discharge requirements. The powders are placed into hoppers and added automatically at a calculated rate with a metering screw conveyer delivery system.

The operation is automated with a programmable automation controller (PAC), eliminating the need for constant operator input. The dissolution occurs in a series of three batch mixing tanks, the timing of which is staggered to contribute to steady flow out of the subsequent dilution tank prior to discharge, where the output borated coolant is supplied to the desired source without the need for an additional outlet pump. A process flow diagram for the system is shown in Figure 3.

Process Flow

The system process can be divided into four main sub-system processes: water inlet and water heating, insulated chamber batching and heating, powder handling and metering, and dilution and discharge.

Water Inlet and Water Heating

Water is provided by an external source with a supply pressure between the ranges of 50 – 150 psig (0.34 – 1.03 MPa) in order to provide the motive force through the system. The inlet connection is designed as an industry standard flange connection to promote universal use. Within the system, flow is split with a portion sent to one of the three dissolution tanks within the insulated chamber while the other portion is provided directly to the inlet of the dilution tank. Prior to entering the insulated chamber, the flow stream is heated to a target temperature using a high-efficiency custom-designed once-through tankless water heating assembly. Flow for both streams is controlled using control valves and proportional-integral (PI) flow meter feedback.

Insulated Chamber Batching and Heating

The outlet flow stream from the tankless water heater enters the insulated chamber and is used to fill one of the three dissolution tanks. Because of the timing aspects demanded from a continuous system, it is imperative that the system heating efficiency and flow rates are maintained.

The three batch-style dissolution tanks are operated in a series consisting of filling, mixing, and draining out of phase in order to provide a continuous overall discharge. During the filling stage, heated water and chemical powders enter the target dissolution tank simultaneously until fill set points are reached. A high concentration solution is targeted in order to improve overall system energy efficiency. Due to the endotherm associated with the dissolution of boric acid (H_3BO_3), band-style heaters are located along the outer circumference of each of the dissolution tanks to add heat to the mixture and maintain a target temperature.

All three dissolution tanks and any process lines that contain high-concentration boric acid solution is located within an insulated chamber designed to maintain an operating temperature above the precipitation limit of the process fluid. Dissolution tank discharge is provided by a dedicated positive displacement pump.

Powder Handling and Metering

The dry powder side of the system consists of two hopper-style holding tanks; one for the boric acid and the other for a pH control agent. The boric acid hopper contains agitation equipment in order to maintain the contents in a free-flowing powder form and prevent agglomeration. The addition to the hopper is accomplished addition to a smaller loading vessel through a series of mechanical auger-style conveyors. The smaller loading vessel is customized to the customer powder container requirements which can range from smaller barrel-style containers to large pallet-sized bulk bag containers. While the powder delivery interface can vary, the goal of transporting the dry powder contents to the larger storage hopper remains the same.

Due to the decreased relative ratio of pH control agent powder use to the boric acid powder, the pH control agent hopper and equipment size demands are much less. The pH control agent is preferred as sodium hydroxide due to boric acid solubility enhancement and exothermic dissolution characteristics. Other pH control agents can be used such as lithium hydroxide. The system can also operate without a pH control agent at a decreased performance profile.

Both the boric acid and the pH control agent are added to the filling dissolution tank using a series of both metering screw conveyors and vertical transport screw conveyors. Since the target concentration of the dissolution tanks is a critical design aspect that has a large effect on the discharge concentration, the metering screw conveyor mass flow rate precision and accuracy were characterized through a series of tests in order to achieve a desired operational performance confidence level. The screw conveyors are all controlled using the PAC sequencing logic for the process.

Dilution and Discharge

After the dissolution tank contents have been added and a mixing cycle is completed, the tank is prepared to discharge to the dilution tank. The dissolution tank discharge is accomplished by the use of positive displacement pumps. Since the tanks are at a relatively high temperature, atmospheric pressure, and minimal elevation head due to the compact system design, a pump that can tolerate low net positive suction head conditions was required. The dissolution tank pumps are operated in direct series with the previously draining tank in order to provide a constant high-concentration discharge flow rate. This is accomplished by having the two pumps communicate through the PAC during de-energization and startup of the two pumps.

The dilution tank receives inlet flow from both the dilute stream and the high concentration stream at calculated and controlled flow rates in order to deliver a known concentration and flow rate at the system discharge based on user inputs.

Supporting Equipment

Supporting equipment includes an air compressor, heating, ventilation, and air conditioning (HVAC) unit, and a motor control center (MCC). The air compressor is used as the energy source for all of the system valves and also serves for a separation medium between the dry powder and fluid system. A HVAC unit is included in the system skid package in order to establish a temperature-controlled environment for both the system operators and equipment. The MCC allows for a compact arrangement to control and monitor all of the system motors.

Design Aspects

System Design

While the system's main function is to generate borated fluid, another important functional requirement is that the system be designed and constructed within a mobile platform. The important characteristics for a mobile system are to allow for a sufficiently rugged design for transport while still maintaining a functional system with minimal setup requirements or constraints. The system was modeled using a computer-aided drafting design interface and acceleration force loadings on components met a customer-specific seismic load as well as an interface standard for highway transportability (Reference 4). Pressure retaining components were also designed to meet ASME B31.1 requirements (Reference 5). Design considerations were also given for the unit center of gravity, weight distribution, and floor loading. Anticipated travel pathway size restrictions limited vertical and horizontal dimensions. The final design was constructed within an International Standards Organization (ISO) high-cube 40' cargo container, a picture of which is shown in Figure 5.

There are some accessibility aspects to the system that require additional physical deployment. The rear of the container has a foldable platform that is used for both operator access and powder staging. Another access point was designed near the front of the container used solely for operator access.

With a mobile system, the system is inherently compact that can make component accessibility and maintainability challenging. While the design container uses a standard ISO container, minor customization was required to enhance accessibility. As an example, side panels were added to be able to access components within the insulated chamber. This allows for easy operator access to all instruments, valves, flanges, and heating equipment within this chamber. This accessibility is shown in Figure 4.

Water and System Heating

Of the allotted 200 – 300 kW power budget to operate the entire system, about 150 – 200 kW are used to heat fluid. With this high percentage of energy devoted to water heating, it is important for the system to use this energy as efficiently as possible. Since the boric acid solubility curve is non-linear, there is more benefit for a higher dissolution tank concentration which in turn requires a higher operating temperature. This relationship is outlined in Figure 6. The figure was developed during the prototype development stage and was using a nominal design output target of 2600 ppm boron and 75 gpm (17 m^3/hr). The system flow rate on the secondary axis corresponds to the dissolution tank discharge flow and considers a subsequent dilution in order to achieve the target flow of 75 gpm (17 m^3/hr). Nevertheless, a few important design trends can be established and were used for the scaled up production model. While the system inlet temperature is varied between 37°F (3°C) and 77°F (25°C), the overall energy requirements decrease. This is an obvious conclusion since more heat is required to raise a higher fluid temperature delta. A more useful trend was that the overall heat power requirement begins a decreasing trend as the system (dissolution tank) temperature increases. Essentially, there is added benefit the higher the system operating temperature. Temperatures beyond 212°F (100°C) will also add thermal benefit, but the design complications and safety issues associated with an elevated pressure high temperature fluid were perceived to outweigh any efficiency benefit. Subsequently, this design concept was not pursued. This water heating is accomplished in two stages; the first stage through a tankless water heating system and the second is through tank band heaters.

The tankless water heater was custom designed using commercially available tankless water heater components in order to meet wetted component requirements commonly associated with nuclear primary water system chemistry. The design of the tankless water heater assembly provided enough resistance downstream from the heating elements to allow for enough pressure to prevent localized boiling near the heating elements. This design extends heater life and stabilizes flow control feedback during dissolution tank fill.

The reaction kinetics and thermodynamics are both considered in the dissolution tank design. Since the overall design requires a continuous discharge, the timing of each dissolution tank has a fixed time interval over which the dissolution tank contents must be added, heated, and dissolved. These requirements create the need for quick and efficient tank heating. Sub-scale testing showed that the overall timing is driven by thermal addition and not dissolution. The addition and dissolution of boric acid in water is an endothermic reaction that can decrease the overall solution temperature by up to 40°F (22°C). The timing requires this temperature be increased back to operating temperature in about 10-12 minutes. These timing and heating functional requirements set the basis for the sizing of the dissolution tank heaters. The high concentration solution prevents a high heat flux submersion heater design; therefore, an external band heater design was used. Multiple band heater designs were investigated with a manufacture that drove towards a custom heater design. Initial sub-scale tests identified local band heater overheating and thermal striping (Figure 1). Multiple design enhancements were made to the tank surface area, band heater power, band heater size, and clamping mechanisms to achieve a functional design.

Figure 1: Band Heater Overheating

Ambient environment heating of the insulated chamber is accomplished using finned strip heaters and temperature feedback control from six resistance temperature detector (RTD) probes at varying locations to obtain a representative temperature. Having a uniform temperature inside the chamber is important to prevent local colder environments. In order to accomplish thermal uniformity, high-temperature fans were included in the chamber. The removable panel design that allows for chamber accessibility was designed to limit thermal loss through a sealed and insulated interface to prevent heat loss.

High Concentration Dilution

The dilution tank design takes the input from two streams at controlled mass flow rates in order to output the desired flow rate and concentration. To achieve this requirement, a basic mass flow around the dilution tank control volume leads to Equation 1.

$$\boldsymbol{M_1 c_1 + M_2 c_2 = M_3 c_3} \qquad \text{Equation 1}$$

Where:

M_1 = Dilute stream inlet mass flow
C_1 = Dilute stream inlet concentration (Equal to 0)
M_2 = High concentration stream inlet mass flow
C_2 = High concentration stream inlet concentration
M_3 = Dilution tank outlet mass flow (Equal to $M_1 + M_2$)
C_3 = Dilution tank outlet concentration

In this sense, the system dilution design is relatively simple. The risk of precipitation associated with the addition of a high temperature, high concentration stream with a colder dilute stream was mitigated by the addition of a compressed air gap between the cooler bulk fluid and high concentration stream. The gap is controlled by the use of a level control system and vent valves in order to prevent gross air entrainment in the dilution tank discharge stream. The precipitation risk is further reduced through the use of automated flushing procedures during normal and off-normal conditions and a manual drain feature in the event of a complete loss of power event.

The mixing of the dilution tank contents is accomplished without mechanical agitation assistance. The dilution tank inlet streams were sized to generate sufficient mixing of the two streams for the range of discharge flow rates and concentration conditions.

Powder Handling

The powder handling and delivery system design used common commercially available powder handling equipment. The unique aspects included the ability to consistently and continuously meter a specified amount of powder as well as be able to continuously add the powder to a humid environment without blockage.

The aspect of adding a metered amount of powder is accomplished through the use of screw conveyors operated with variable frequency drives (VFD). The VFD speed and control was characterized with a series of full-scale tests with the output achieving repeatability confidence level within overall system accuracy targets. The final system characteristics were established as operational constants within the PAC interface.

The powder agglomeration prevention mechanisms were required to be continuous without the use of any operator actions. In order to achieve these requirements, two measures were included in the design. The first was to provide a dry compressed air injection pathway in line with the powder delivery system at the dissolution tank interface in order to prevent any moisture from travelling from the warm, humid environment in the dissolution tank. The actuation of this compressed air path was controlled automatically to be active whenever powder was being added. The second mechanism was to provide the ability for a hot water flush of the components and pipeline downstream from the dissolution tank isolation mechanism. Since the tank envelope will essentially always have a humid environment, there is always a collection area present. An active wash after powder addition was a method tested to allow for a complete wash of any powder accumulation areas.

Since the dissolution tanks must operate at atmospheric pressure and level indication is provided by a pressure measurement near the bottom of each tank, it was imperative to keep pressure communication between the tanks and atmosphere that would have been eliminated if powder blockage or powder buildup occurred in any of the tank vent lines. The tank vent lines were needed in order to direct any entrained powder dust away from the component environment. These lines are monitored during operation, have the ability to be cleaned on-line, and are automatically flushed after each use.

Programmable Automation Controller Design

The PAC design incorporates logical functions in the format of ladder logic, sequential functional charts, and functional block diagrams using a commercially available controller. The interface of which is included as a touch-screen format that displays common system-level graphical information about the system such as operating mode, equipment state, instrument feedback, and alarms.

The interface requires operator inputs in order to initiate the startup sequence. These inputs are operating mode (with or without pH control agent), target concentration, and flow rate. The system performance follows a curve between these two inputs as outlined in Figure 2.

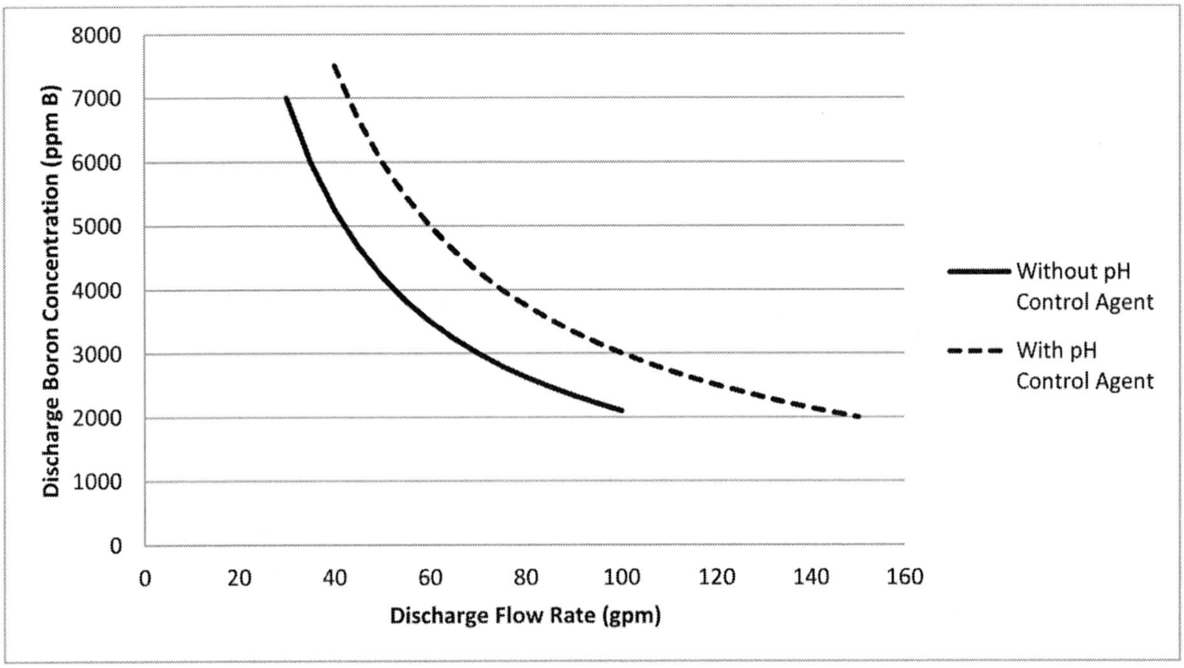

Figure 2: System Performance Curve

Once the inputs are received, the PAC sets target flow rates for the system and the system is initiated. The startup sequence initializes in the following order: insulated chamber heating and fill, powder initiation and transport to dissolution tanks, dilution tank fill, approach to steady state, and steady state established. Once steady state is established, the system is operating at the target conditions.

System Testing

Multiple sub-scale and equipment-level test programs were conducted prior to initiating the full scale, system level test program. During the system level test, the acceptance criteria included both meeting the target flow rate and target concentration within an acceptable error band. Three target conditions were tested. These conditions and results are presented in

Table 1.

Table 1: Mobile Boration System Test Results

Test	Target	Results (Note 1)
Test 1	7000 ppm-B; 30 gpm; No pH agent	Sample 1: 6913 ppm-B (Titration); 7390 ppm-B (ICP-AES) Sample 2: 6987 ppm-B (Titration); 7200 ppm-B (ICP-AES) Sample 3: 6946 ppm-B (Titration); 6880 ppm-B (ICP-AES) Steady state average flow rate about 30 gpm
Test 2	7000 ppm-B; 30 gpm; No pH agent	6933 ppm-B (Titration); 6884 ppm-B (Titration); Steady state average flow rate about 30 gpm
Test 3	3000 ppm-B; 100 gpm; pH agent	Sample 1: 3090 ppm-B (ICP-AES) Sample 2: 2930 ppm-B (ICP-AES) Steady state average flow rate about 100 gpm
Test 4	3000 ppm-B; 100 gpm; and on-line system change to 2000 ppm-B; 150 gpm; pH agent	Sample 1 (3000 ppm-B target): 3200 ppm-B (ICP-AES) Sample 2 (2000 ppm-B target): 2160 ppm-B (ICP-AES) Steady state average flow rate of about 100 gpm and 150 gpm as required per target

Note 1: Multiple test sample results are extractions at different test durations. The test method is included in parenthesis.

Along with testing the functional requirements of the system, the entire PAC system architecture was testing, which included human-machine interface (HMI) navigation, alarm function, control strategy stability, and margin assessments.

Enhancement of Plant Safety

In the emergency situations addressed by a specific regulator, it may be difficult to have ready access to borated fluid if current systems or storage tanks are not designed to withstand the postulated severe external event scenarios. To date, normal, emergency or augmented boric acid supply systems have relied on the existing single-batch tank systems because of the system requirements of continuously dissolving boric acid crystals in water, which takes a significant amount of energy for heating. This required high-energy heating and dissolution process limits the overall system capacity for a single-batch system. This system capacity can be directly translated to an overall discharge flow rate of borated fluid, which may be too low to meet the high flows necessary to alleviate early post-accident decay heat rates.

Within the United States, Nuclear Regulatory Commission (NRC) Order EA-12-049 (Reference 6) was developed and the Nuclear Energy Institute published the subsequently NRC-endorsed NEI 12-06 (Reference 7). The FLEX guide outlines requirements related to core cooling heat removal and a method of replenishing borated water for pressurized water reactor coolant systems indefinitely.

Requirements for addressing core cooling and heat removal for Modes 5 and 6, cold shutdown and refueling, respectively, as outlined in the FLEX guide, stipulate that all plants have a means to provide borated water to replenish the reactor coolant system. During these modes, the normal methods for core cooling are assumed to be inoperable. The reactor coolant system is designed to vent heat through steam into containment as part of the core cooling process, but the boric acid from the borated fluid within the system will remain as precipitate. A flow of borated water must be available to keep the boric acid concentration below precipitation limits and remove decay heat.

The continuous flow mobile boration system, designed, built, and tested as discussed, is a unique solution for plants that require an indefinite protected or diverse borated water source.

Operational Efficiency

Existing systems in plants today use a single batching tank to replenish the reactor coolant system and all of a plant's fluid systems that require borated fluid. This approach is a manual, operator-intensive process that can be time-consuming and cannot provide continuous flow. Typically, these existing systems consist of an unenclosed batching tank to which water and boric acid powder are added and the fine boric acid power dust can escape to the surrounding environment. Logistically, moving the powder containers around sites to replenish the various systems often means travelling through protected areas and can involve additional security precautions. The overall time to batch and add the amount of borated water inventory required for normal operation can take upwards of a week.

Summary

Following the Fukushima Daiichi event, Westinghouse responded to international industry interest in a skid-style mobile boration system as part of a mitigation strategy for core cooling that would address international regulatory requirements. The finalization of this design was able to provide that the functional requirements for such an event were able to be met and was proven through a commercial system. Future applications can include both the enhancement of plant safety and the improvement of operational efficiency at a nuclear power plant.

Nomenclature

FLEX Flexible and Diverse Coping Strategies

HMI human-machine interface

HVAC heating, ventilation, and air conditioning

ISO International Organization for Standardization

MBS mobile boration system

MCC motor control center

NEI Nuclear Energy Institute

NRC Nuclear Regulatory Commission

PAC programmable automation controller

PI proportion integral

RTD resistance temperature detector

VFD variable frequency drive

References

1. United States Patent Number 9,266,756, "Mobile Boration System," February 23, 2016.
2. United States Patent Number 9,281,087, "Mobile Boration System," March 8, 2016.
3. LePree, J. (2015). Moving to Modular. *Chemical Engineering*, 16-18.
4. MIL-STD-1366E, "Interface Standard for Transportability Criteria," Department of Defense, October 2006.
5. ASME B31.1-2014, "Power Piping," ASME Code for Pressure Piping, B31.
6. NRC Order EA-12-049, "Order Modifying Licenses With Regard To Requirements For Mitigation Strategies For Beyond-Design-Basis External Events," March 12, 2012, NRC ADAMS Accession Number ML12056A045.
7. NEI 12-06, Revision 3, "Diverse and Flexible Coping Strategies (FLEX) Implementation Guide", September 2016, NRC ADAMS Accession Number ML16267A274.

ANNEX A

MOBILE BORATION SYSTEM FIGURES

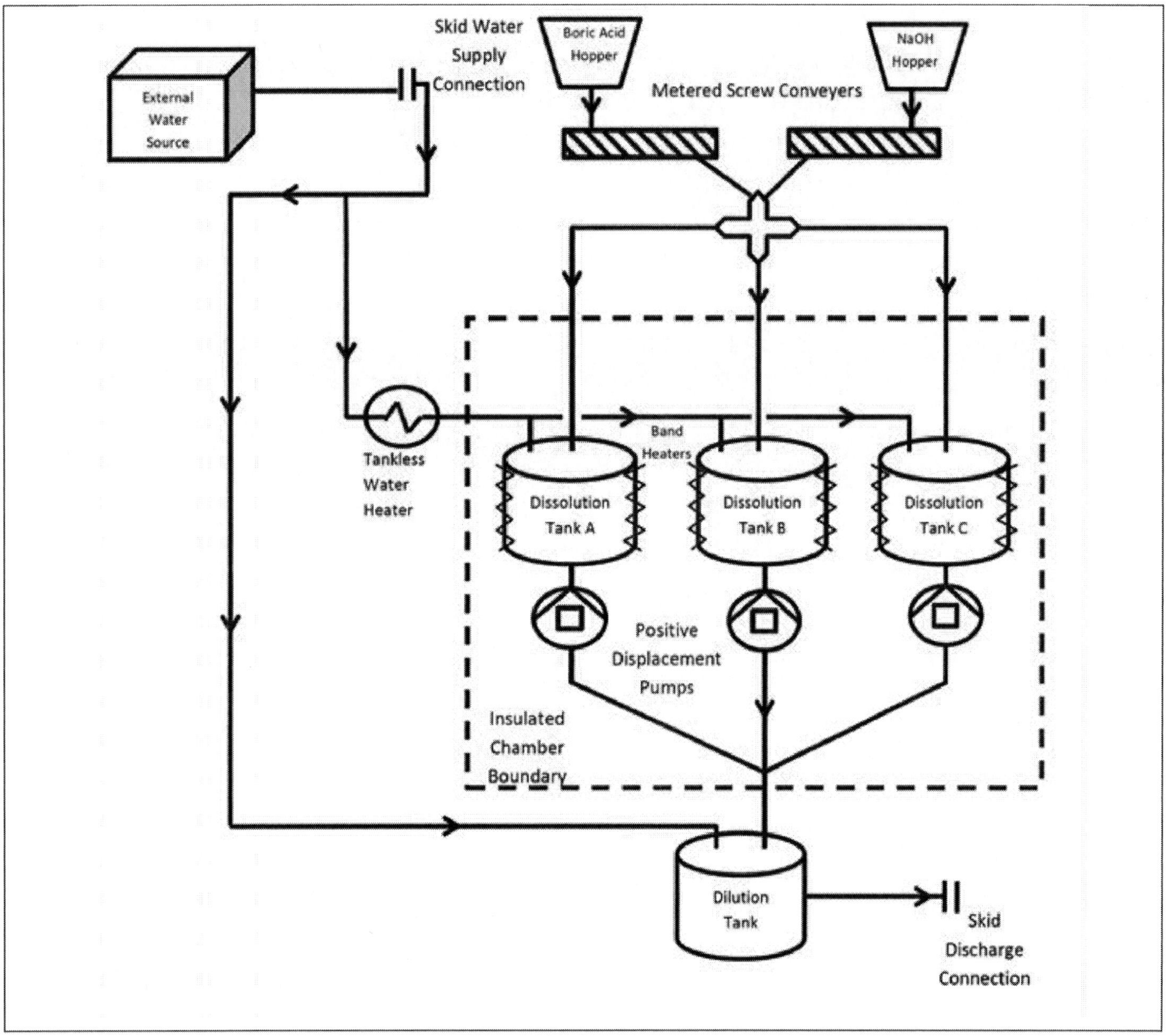

Figure 3: MBS Process Flow Diagram

Figure 4: Insulated Chamber Access Panels

Figure 5: External System

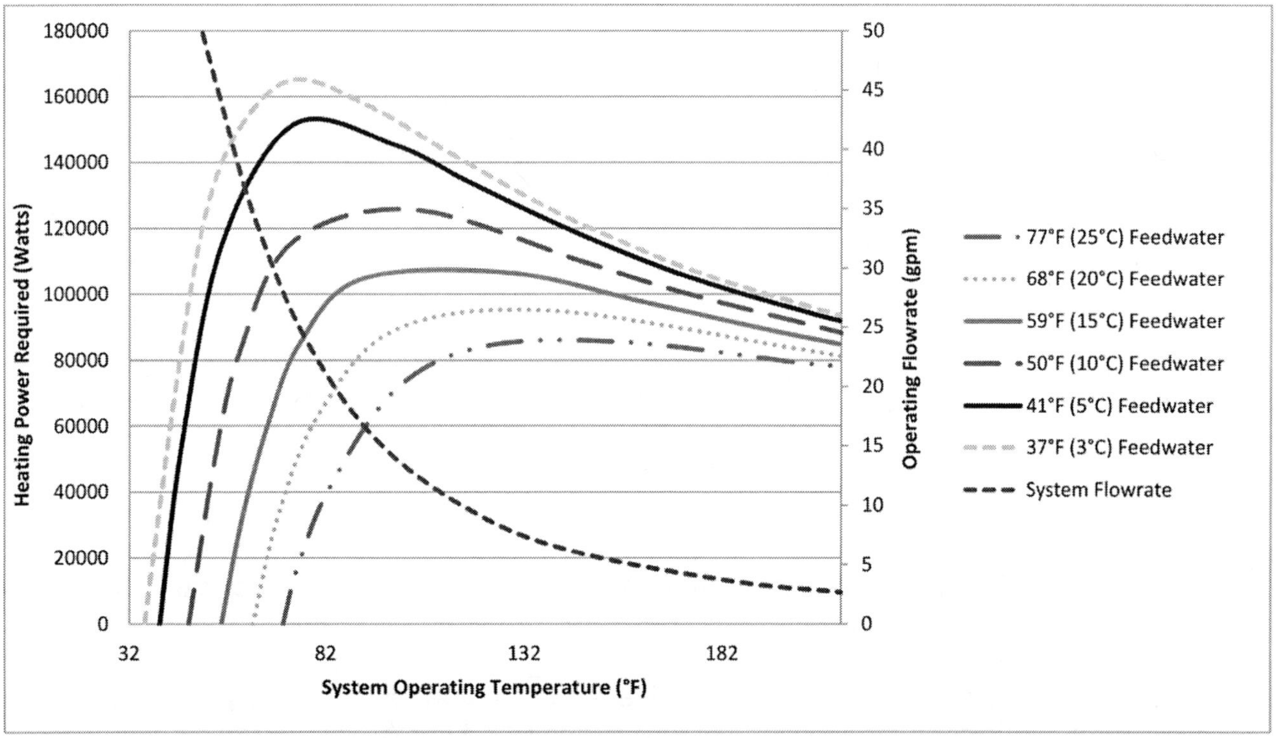

Figure 6: System Operating Temperature and Overall Power Requirements

ANALYSIS OF AIR OXIDATION OF PLUTONIUM METAL

James E. Laurinat
Savannah River National Laboratory
Aiken, South Carolina

Abstract

Rates of oxidation of plutonium metal alloys exposed to high-temperature air are analyzed. Department of Energy facilities, including the Savannah River Site, utilize direct metal oxidation in air (DMO) to convert surplus plutonium metal alloys to a stable oxide form. The plutonium oxide can be either stored or used as feed for other processes. Typically, DMO is performed in an electrically heated furnace; the furnace supplies the heat needed to raise the metal temperature to its ignition point, which is a requirement for oxidation in dry air.

Oxidation rates are calculated for oxidation of a 3.5-in. diameter by 1.25-in. thick cylinder and for a 4-in. by 4-in. by 0.125-in. piece of foil. The calculations demonstrate that the oxidation rates are not limited by the build-up of the oxide powder layer on the surface of the metal but instead are kinetically controlled. Heat transfer calculations show that the air supply to the furnace must be throttled to limit the furnace temperature after ignition.

Separate analyses are included to estimate the minimum oxygen concentration in the air and the porosity of the oxide powder generated by the oxidation. The minimum oxygen concentration is estimated from the likelihood that an oxygen molecule will diffuse to an oxidation reaction site on the metal surface. The porosity is calculated from the assumption that the oxide layer expands to a minimal extent in the direction normal to the metal surface.

Introduction

Plutonium metal oxidation rates are analyzed for the high temperature air oxidation of plutonium metal pieces. Rates are analyzed for both alpha phase and stabilized delta phase plutonium metal. Oxidation times are calculated for two representative pieces, a 0.0889-m diameter by 0.0318-m thick cylindrical "button" and a 0.1016-m by 0.1016-m by 0.00318-m foil.

The analysis of plutonium burning rates is based on information obtained from a literature search and a review of Savannah River Site operating experience. Savannah River Site tests have shown that plutonium can be oxidized either by moist air at approximately 473 K or by dry air at or slightly above the plutonium ignition temperature, which is about 773 K.

During moist air oxidation, the hydrogen from the water vapor forms a plutonium hydride phase, which then readily oxidizes. The moist air oxidation method was used to recycle plutonium with excessive impurity levels in the Savannah River Site FB-Line. This method avoided any risk of raising the temperature past the plutonium metal melting point of 913 K.[1] (Molten plutonium metal alloys with stainless steel and other metals such as tantalum; the formation of such alloys was undesirable for plutonium production.) The rate for moist air oxidation was sufficient to convert a plutonium button to oxide within one day, provided that the plutonium metal was alpha phase. Delta-stabilized plutonium is not susceptible to attack from water vapor and therefore oxidizes unacceptably slowly.

High temperature oxidation (also known as direct metal oxidation or DMO) involves direct attack of the metal surface by oxygen. If the plutonium surface temperature exceeds approximately 743 K (or 773 K for delta-stabilized plutonium),[2] the metal ignites. After ignition, the metal surface temperature undergoes an excursion to a temperature that can exceed 1073 K, before dropping to an equilibrium level that depends on the rate of cooling of the metal piece. The oxidation reaction is highly exothermic, so no heat source is required to sustain the reaction past the ignition point, i.e., the reaction becomes autothermic. Experience has demonstrated that it is difficult or impossible to control the metal surface temperature to prevent ignition in a processing environment. Typically, instead, the air feed rate is limited to control the rate of oxidation so that heat can be removed at the rate it is generated by the reaction. In this calculation the DMO oxidation approach is evaluated.

Los Alamos National Laboratory has conducted DMO tests with the objective of converting plutonium from weapons components into oxide.[3] In the Los Alamos DMO process, the plutonium metal is oxidized while being tumbled in a wire basket. The tumbling removes the outer oxide layer to improve heat transfer from and mass transfer to the oxidizing surface. Tests were conducted using a 1:2 helium/oxygen mixture at temperatures ranging from 748 K to 873 K.

At temperatures below the ignition point, the oxidation rate varies with temperature as predicted by an Arrhenius curve but undergoes several step changes that generally coincide with metal phase transitions. In particular, after the plutonium metal transitions to the delta phase, at approximately 673 K, the oxidation rate drops as a protective layer of substoichiometric oxide forms at the metal surface. The oxidation rate then rapidly climbs until ignition is reached. After ignition, the intrinsic chemical oxidation reaction is limited by solid phase diffusion of negatively charged oxygen ions across the substoichiometric surface layer.[4] The intrinsic oxidation rate is constant with temperature up to about 1273 K.[5] The overall oxidation rate for a metal surface covered by a thick oxide layer may be limited by gas phase diffusion through the oxide layer.

The following two sections present analyses of the minimum oxygen concentration required for ignition of plutonium metal and the effect of the oxide layer thickness on the rate of oxidation. The latter analysis shows that the maximum effective oxide layer thicknesses for the foil and the cylindrical "button" do not reach the critical thickness at which the intrinsic oxidation rate is inhibited. Consequently, the oxidation rates for both pieces can be considered to be kinetically controlled. The last section of this paper presents a heat transfer analysis, conducted to determine the maximum reaction rate at which the temperature of the oxide inside a furnace can be limited so that it does not significantly exceed the ignition temperature.

Analysis of Threshold Oxygen Concentrations for Ignition of Plutonium Metal

The calculation of the threshold volume fraction oxygen required for ignition of a plutonium metal object is based on the number of available, i.e., adjacent, molecules at each oxidation reaction site. The theory behind this calculation is that, for the oxidation reaction to propagate through the metal, the probability of an oxygen molecule contacting and reacting at a given reaction site must be greater than 0.5. If more than 50% of the available reaction sites are active, the number of active sites will crowd out the number of inactive sites and the oxidized layer will propagate through the metal. If fewer than 50% of the sites are active, the inactive sites will predominate and the oxidation will fail to proceed.

A simplifying assumption is made that the air molecules surrounding each reaction site are arranged at a uniform density as close-packed, uniform size spheres. The number of available molecules at each equivalent molecular radial distance from the reaction site is assumed to be equal to the number of tangent spheres in that shell layer within the close packed arrangement. For the first few shell layers, the number of tangent spheres contains a discretization error. To minimize this error, the equivalent number of tangent spheres is calculated for successive shells proceeding outward until this number converges. The equivalent number of tangent spheres is computed by normalizing the gross count by the square of the equivalent radius for that shell.

For the initial sphere, there are three spheres located on the plane above and six spheres in the same plane. The three spheres in the plane below are not counted because they are assumed to lie below the surface. Only half the number of spheres in the surface plane is counted, giving a total of six spheres in contact. The second shell contains 12 spheres in the same plane, nine spheres in the first layer above the plane, and either six or seven spheres in the second layer above the plane, depending on the second layer orientation with respect to the first layer. The total number of spheres in the second shell is half of 12, or six, plus nine plus 6.5, for a total of 21.5. A similar count can be performed for each succeeding shell. The following equations summarize the number of spheres in each succeeding shell:

$$n_1 = 6 \tag{1}$$

$$n_2 = 21.5 \tag{2}$$

$$n_i = 2n_{i-1} - n_{i-2} + 10.5,\ i \geq 2 \tag{3}$$

where

n_1 = number of spheres in the first shell

n_2 = number of spheres in the second shell

n_i = number of spheres in the ith shell

n_{i-1} = number of spheres in the shell one layer inside the ith shell

n_{i-2} = number of spheres in the shell two layers inside the ith shell

The effective number of tangent spheres is computed by multiplying the gross number of tangent spheres by the ratio of the shell radius to the radius of the first shell, which is two sphere radii:

$$n_{i,eff} = n_i \left(\frac{r_1}{r_i} \right)^2 \tag{4}$$

where

$n_{i,eff}$ = equivalent number of spheres in the ith shell, projected to the first shell

r_i = radius of the ith shell

r_1 = radius of the first shell

The shells are not propagated uniformly, but by layer. Therefore, the effective radius in the plane tangent to the reacting surface is greater than the effective radius in the normal plane. The average radius is calculated as a weighted geometric mean of the tangential and normal radii. The tangential radius is weighted twice as heavily as the normal radius to account for the fact that the tangential radius expands in two directions, compared to one direction for the normal radius. The average radius is given by:

$$r_i = \left(r_{i,tan}\right)^{2/3} \left(r_{i,norm}\right)^{1/3} \tag{5}$$

where

$r_{i,tan}$ = average radius of the ith shell, tangent to the plane of the reacting surface

$r_{i,norm}$ = average radius of the ith shell, normal to the plane of the reacting surface

The ratio of the tangential radius to the nominal radius is equal to the square root of the ratio of the areas of a circle and a hexagon inscribed in that circle. Thus,

$$r_{i,tan} = ir_1\left(\frac{3(3)^{0.5}}{2\pi}\right)^{0.5} = 0.9094ir_1 \tag{6}$$

The ratio of the normal radius to the nominal radius equals the ratio of the close packed sphere layer height to the sphere radius. Therefore,

$$r_{i,norm} = ir_1\left(\frac{(6)^{0.5}}{3}\right) = 0.8165ir_1 \tag{7}$$

The number of molecules adjacent to each reaction site is evaluated as the limiting number of tangent spheres as the number of shell layers expands:

$$n_{eff} = \lim_{i\to\infty}\left(n_{i,eff}\right) = 6.82 \tag{8}$$

where

n_{eff} = limiting equivalent number of adjacent spheres, projected to the first shell

The limiting number of adjacent molecules converges on 6.82.

The probability that an adjacent molecule will react is assumed to be the probability that that molecule is an oxygen molecule; in other words, it is assumed that an adjacent oxygen molecule always will react. The probability is computed under the assumption that movement of air toward the surface occurs much faster than the rate of reaction, so that oxygen molecules are replaced by air molecules as they react. It is convenient to visualize the probability as the sum of this oxygen likelihood with replacement for each adjacent molecule. Viewed this way, the overall probability for an oxygen molecule reacting is given by:

$$p(x_s) = n_{eff}\sum_{j=1}^{\infty} x_s^{\,j} \tag{9}$$

where

x_s = oxygen concentration at the surface

$p(x_s)$ = probability of reaction at a reaction site

The series $\sum_{j=1}^{\infty} x_s^{\,j}$ is equal to $\frac{x_s}{1-x_s}$. Thus,

$$p(x_s) = \frac{n_{eff}x_s}{1-x_s} \tag{10}$$

For the oxidation reaction to propagate downward from the surface, the probability of reaction at any given site must be greater than 50%. For the limiting oxygen concentration required for ignition, then,

$$p\left(x_{s,min}\right) = 0.5 \tag{11}$$

where

$x_{s,min}$ = limiting oxygen concentration at the surface

$p(x_{x,min})$ = limiting probability of reaction at a reaction site

The preceding two equations combine to yield:

$$x_{s,min} = \frac{1}{2n_{eff} + 1} \quad (12)$$

The effective number of adjacent molecules is 6.82. For this value,

$$x_{s,min} = 0.0683 \quad (13)$$

The limiting surface concentration is the threshold concentration for ignition for bare plutonium metal in the absence of an oxide layer. If an oxide layer has formed, the minimum oxygen concentration increases due to diffusion through this layer. The minimum bulk oxygen concentration outside the oxide layer can be estimated as a function of the minimum surface concentration through an analysis of the diffusion equation.

Diffusion of oxygen through the oxide layer involves countercurrent diffusion of the nitrogen in air. Customarily, as in evaporation from a liquid surface, this is analyzed by adding a bulk flow term and setting the total mass flux of the noncondensable gas, comprised of diffusive and bulk flow, to zero. Diffusion to a reacting surface differs from evaporation in that the surface reaction rate contains local variations, whereas the evaporation rate from a liquid is nearly uniform. The variations in the local reaction rate can be pictured as setting up recirculation patterns within the oxide layer, where air flows toward the surface and oxygen diffuses to the surface over active reaction zones, and there is a counterbalancing flow of air away from the surface over inactive portions of the surface. The air circulation within the oxide layer is modeled using a dispersion term. With the addition of this dispersion term, the steady state mass transfer equation for flow of air and diffusion of oxygen toward the surface takes the form:

$$v\frac{dx}{dz} = D_{eff}\frac{d^2x}{dz^2} \quad (14)$$

where

z = distance from the oxidizing surface

v = convection velocity in the porous oxide layer

x = mole fraction oxygen

D_{eff} = effective dispersion coefficient for oxygen flow through the oxide layer

The rate of diffusion can be assumed to be sufficiently slow that the convection velocity remains constant throughout the oxide layer. For a constant velocity, the mass transfer equation becomes linear and can readily be solved with the appropriate boundary conditions.

The boundary conditions for the mass transfer equation are:

$$x = x_s \text{ at } z = 0 \quad (15)$$

and

$$x = x_b \text{ at } z = L \quad (16)$$

where

x_b = bulk concentration of oxygen in the gas outside the oxide layer

L = oxide layer thickness

For these boundary conditions, the solution of the mass transfer equation is:

$$x = \frac{\left(\exp\left(\frac{vL}{D_{eff}}\right) - \exp\left(\frac{vz}{D_{eff}}\right)\right)x_s + \left(\exp\left(\frac{vz}{D_{eff}}\right) - 1\right)x_b}{\exp\left(\frac{vL}{D_{eff}}\right) - 1} \tag{17}$$

The fact that the surface oxidation reaction is heterogeneous allows the mass transfer equation to be integrated without the addition of any constants of integration. Because the surface reaction drives the diffusion, the integral mass transfer equation is applied at the metal surface. The surface integral mass transfer equation is:

$$vx_s = D_{eff} \left.\frac{dx}{dz}\right|_{z=0} \tag{18}$$

The solution to this integral equation gives the surface oxygen concentration as a function of the bulk concentration:

$$x_s = \frac{x_b}{2 - \exp\left(\frac{vL}{D_{eff}}\right)} \tag{19}$$

For the remainder of this analysis, it is convenient to introduce a dimensionless Péclet number defined by:

$$Pe == \frac{(-v)L}{D_{eff}} \tag{20}$$

and a concentration ratio defined by:

$$y = \frac{x_s}{x_b} \tag{21}$$

where

Pe = Péclet number for diffusion of oxygen through the oxide layer

y = ratio of the oxygen concentration at the metal surface to the oxygen concentration at the top of the oxide layer

The solution to the integral equation in terms of these dimensionless quantities becomes:

$$y = \frac{1}{2 - \exp(-Pe)} \tag{22}$$

The value of the Peclet number and the solution for the bulk oxygen concentration as a function of the surface concentration is obtained by analyzing the local rate of surface oxidation. The total local rate of reaction $f_R R_{diff}$ equals the sum of the contributions from diffusion and convection:

$$f_R R_{diff} = -vx_s - D_{eff} \left.\frac{dx}{dz}\right|_{z=0} \tag{23}$$

where

R = nominal rate of surface oxidation

f_R = multiplier to account for variations in the rate of surface oxidation

From the integral mass balance, the diffusive and convective mass fluxes at the metal surface are equal. Thus, from this mass balance:

$$f_R R_{diff} = \frac{2Pe}{2-\exp(-Pe)}\left(\frac{D_{eff} x_b}{L}\right) \quad (24)$$

It may be noted that no restrictions have been placed on the surface rate of reaction other than those for diffusion through the oxide layer. In the absence of any such restrictions, the local oxidation rate scales proportionally as the rate of oxide layer diffusion. Linear scaling implies that:

$$\frac{d(f_R R_{diff})}{d\left(\frac{D_{eff} x_b}{L}\right)} = 1 \quad (25)$$

This scaling relationship is satisfied if:

$$\frac{2Pe}{2-\exp(-Pe)} = 1 \quad (26)$$

The iterative solution to this equation takes the form:

$$Pe = Pe + \frac{2-2Pe-\exp(-Pe)}{2-\exp(-Pe)} \quad (27)$$

The solution for the Péclet number and the concentration ratio y is

$$Pe = 0.7680 \quad (28)$$

and

$$y = 0.6510 \quad (29)$$

If an oxide layer has formed, the minimum bulk concentration required for ignition, i.e, the minimum concentration at the top of the oxide layer, is given as a function of the minimum surface concentration by:

$$x_{b,min} = \frac{x_{s,min}}{y} \quad (30)$$

where

$x_{b,min}$ = limiting oxygen concentration at the top of the oxide layer

The solution for the minimum bulk concentration is:

$$x_{b,min} = 0.1049 \quad (31)$$

The calculated minimum surface concentration of 6.8 vol % is in approximate agreement with the measured minimum concentration for ignition of 6 vol %, and the calculated minimum bulk concentration of 10.5 vol % is consistent with the observed concentration of 10 vol % for the onset of accelerated quenching.[2]

Analysis of Limiting Oxide Layer Thickness for Reduction of the Plutonium Oxidation Rate

The minimum oxide layer thickness at which the plutonium metal surface oxidation rate begins to decrease can be estimated by comparing the pressure drop for flow across the oxide layer to the pressure change generated by the oxidation reaction at the limiting condition for ignition. The pressure change created by the reaction is equal to the product of the ambient pressure and the oxygen concentration change across the oxide layer, expressed as a volume fraction. When the oxide layer thickness begins to affect the oxidation rate, this pressure change equals the pressure drop for flow:

$$\Delta P = P_a\left(x_{b,min} - x_{s,min}\right) \tag{32}$$

where

ΔP = pressure drop for flow of oxygen associated with plutonium oxidation across the oxide layer

The pressure drop for flow is given by the Ergun equation, which takes the form:[6]

$$\frac{\Delta P}{L} = \frac{150\mu v\left(1-\varepsilon_{ox}\right)^2}{\varepsilon_{ox}{}^3 d_p{}^2} + \frac{1.75\rho v^2\left(1-\varepsilon_{ox}\right)}{\varepsilon_{ox}{}^3 d_p} \tag{33}$$

where

μ = air viscosity

d_p = average oxide layer particle diameter

ρ = air density

The preceding two equations combine to give an expression for the limiting oxide layer thickness required for a reduction in the rate of oxidation:

$$L_{min} = \frac{P_a\left(x_{b,min} - x_{s,min}\right)}{\dfrac{150\mu v\left(1-\varepsilon_{ox}\right)^2}{\varepsilon_{ox}{}^3 d_p{}^2} + \dfrac{1.75\rho v^2\left(1-\varepsilon_{ox}\right)}{\varepsilon_{ox}{}^3 d_p}} \tag{34}$$

where

L_{min} = minimum oxide layer thickness resulting in a reduction in the rate of plutonium oxidation

For oxide layers thicker than this minimum thickness, the time-average oxidation rate decreases proportionally as the thickness increases.

It is assumed that the bulk flow rate equals the rate of consumption of oxygen at the metal surface. Therefore, the velocity is given by:

$$v = \frac{R_{rxn}}{M_{Pu}\rho_{g,m}} \tag{35}$$

where

M_{Pu} = atomic mass of plutonium, 0.239 kg/mol

$\rho_{g,m}$ = molar gas density

The molar gas density is calculated using the ideal gas relation:

$$\rho_{g,m} = \frac{P_a}{R_g T} \tag{36}$$

where

R_g = gas law constant

The evaluation of the oxide layer porosity is based on the assumption that the volumetric expansion that accompanies the metal oxidation is confined to the direction perpendicular to the oxidizing surface. Moreover, the expansion in this direction is limited to the direction outward from the metal surface. Mathematically, this implies that that the ratio of the void volume to the solid volume within the oxide equals the fraction of the volumetric expansion prevented by this steric hindrance, $1-f$, divided by the total oxide volume associated with the fractional expansion allowed, f. In other words:

$$\frac{\varepsilon_{ox}}{1-\varepsilon_{ox}}=\frac{(1-f)\left(V_{PuO_2}-V_{Pu}\right)}{fV_{PuO_2}} \tag{37}$$

where

ε_{ox} = the oxide layer porosity

V_{PuO_2} = the molar volume of PuO_2

V_{Pu} = the molar volume of Pu metal

To calculate the porosity, the fraction f and the molar volumes must be evaluated. The ratio of the two molar volumes is given by:

$$\frac{V_{PuO_2}}{V_{Pu}}=\frac{M_{PuO_2}\rho_{Pu}}{M_{Pu}\rho_{PuO_2}} \tag{38}$$

where

M_{PuO_2} = molecular mass of PuO_2, 0.271 kg/mol

The fraction of the expansion that is not prevented by steric hindrance can be calculated by integrating the outward normal from the oxidizing surface over a uniform distribution of vectors, which takes the shape of a sphere. Thus, if θ is the angle between a random vector and the outward normal vector, then:

$$f=\frac{\int_{-\frac{\pi}{2}}^{0} r^2\sin^2(\theta)\cos(\theta)\,dr\cos(\theta)}{\int_{-\pi}^{0} \pi r^2\sin^2(\theta)\,dr\cos(\theta)} \tag{39}$$

where

r = radial distance from the center of a sphere

θ = angle between random vector and outward normal vector from the center of a sphere

Integration of this equation gives:

$$f=\frac{3}{16} \tag{40}$$

Substitution of Equations 38 and 40 in Equation 37 yields:

$$\varepsilon_{ox}=\frac{13\left(\frac{M_{PuO_2}\rho_{Pu}}{M_{Pu}\rho_{PuO_2}}\right)-13}{16\left(\frac{M_{PuO_2}\rho_{Pu}}{M_{Pu}\rho_{PuO_2}}\right)-13} \tag{41}$$

The oxide layer porosity is computed to be 0.627.

From the preceding analysis, the minimum oxide layer thickness required for a reduction in the rate of oxidation after ignition is 0.0891 m for an oxidation rate of 1.0 kg $Pu/m^2/min$, 0.0677 m for an oxidation rate of 1.4 kg $Pu/m^2/min$, and 0.0424 m for an oxidation rate of 2.1 kg $Pu/m^2/min$.

The oxide layer thickness is calculated from the following volume balance, which equates the ratio of the oxide layer volume for the oxidizing specimen to the theoretical ratio of oxide to metal volumes from Equation 41:

$$\frac{V_{tot} - V_{Pu}}{V_{Pu,0} - V_{Pu}} = \frac{\rho_{Pu} M_{PuO_2}}{\rho_{PuO_2} M_{Pu} \left(1 - \varepsilon_{ox}\right)} \tag{42}$$

where

V_{tot} = total volume of the remaining plutonium metal and oxide layer

V_{Pu} = volume of the remaining plutonium metal

$V_{Pu,0}$ = initial volume of the plutonium metal piece prior to oxidation

The oxide layer thickness is evaluated for the two pieces considered in this report, the 0.0889-m diameter by 0.0318-m high cylinder and the 0.1016-m by 0.1016-m by 0.00318-m square foil. If the top and side surfaces of the cylinder are assumed to oxidize at a uniform rate, then the volumes of the remaining metal, the original metal cylinder, and the total volume including the metal and the oxide layer are given by:

$$V_{Pu} = \frac{\pi \left(d_{m,0} - 2\Delta t_m\right)^2 \left(t_{m,0} - \Delta t_m\right)}{4} \tag{43}$$

$$V_{Pu,0} = \frac{\pi d_{m,0}{}^2 t_{m,0}}{4} \tag{44}$$

and

$$V_{tot} = \frac{\pi \left(d_{m,0} + 2\Delta t_{tot}\right)^2 \left(t_{m,0} + \Delta t_{tot}\right)}{4} \tag{45}$$

where

Δt_m = change from the original thickness of the plutonium metal piece

For the square foil, the corresponding volumes are:

$$V_{Pu} = \left(s_{m,0} - 2\Delta t_m\right)^2 \left(t_{m,0} - \Delta t_m\right) \tag{46}$$

$$V_{Pu,0} = s_{m,0}{}^2 t_{m,0} \tag{47}$$

and

$$V_{tot} = \left(s_{m,0} + 2\Delta t_{tot}\right)^2 \left(t_{m,0} + \Delta t_{tot}\right) \tag{48}$$

For both the cylinder and the foil, the oxide layer thickness is equal to the sum of the changes in the metal piece and the total thicknesses:

$$L = \Delta t_m + \Delta t_{tot} \tag{49}$$

A corrected oxide layer thickness can be estimated by accounting for dispersion of the oxygen flow across the layer. The dispersion follows the same pattern as the dispersion of the heat flux across an insulating layer, for which the effective thickness is proportional to the actual thickness times the square root of the ratio of the inner and outer surface areas. Thus,

$$L_{corr} = \left(\frac{A_m}{A_{tot}}\right)^{0.5} \left(\Delta t_m + \Delta t_{tot}\right) \tag{50}$$

where

L_{corr} = oxide layer thickness, corrected for the effect of gas dispersion due to the change in cross-sectional area as the oxide layer grows

A_m = surface area of the remaining metal piece available for oxidation

A_{tot} = outer surface area of the oxide layer

The metal piece and total surface areas for the cylinder are:

$$A_m = \frac{\pi\left(d_{m.0} - 2\Delta t_m\right)^2}{4} + \pi\left(d_{m,0} - 2\Delta t_m\right)\left(t_{m,0} - \Delta t_m\right) \tag{51}$$

and

$$A_{tot} = \frac{\pi\left(d_{m.0} + 2\Delta t_{tot}\right)^2}{4} + \pi\left(d_{m,0} + 2\Delta t_{tot}\right)\left(t_{m,0} + \Delta t_{tot}\right) \tag{52}$$

For the foil,

$$A_m = \left(s_{m.0} - 2\Delta t_m\right)^2 + 4\left(s_{m,0} - 2\Delta t_m\right)\left(t_{m,0} - \Delta t_m\right) \tag{53}$$

and

$$A_{tot} = \left(s_{m.0} + 2\Delta t_{tot}\right)^2 + 4\left(s_{m,0} + 2\Delta t_{tot}\right)\left(t_{m,0} + \Delta t_{tot}\right) \tag{54}$$

Figure 1 depicts the variation of the oxide layer thickness during oxidation of the 0.0889-m diameter by 0.0318-m high cylinder. Without any correction for dispersion of the oxygen flow, the maximum oxide layer thickness is 0.0568 m, at the end of the oxidation transient. With the dispersion correction, the effective oxide layer thickness peaks at 0.0147 m, when approximately half of the piece has oxidized. Figure 2 portrays the variation of the oxide layer thickness for oxidation of the 0.1016-m square by 0.00318-m thick rectangular prism. The maximum oxide layer thickness for this piece is 0.0106 m without the correction for dispersion and 0.0075 m with this correction. The maximum effective oxide layer thicknesses are significantly less than the minimum thickness at which the oxidation rate is reduced. Therefore, it may be stated that the oxide layer does not inhibit oxidation for the two metal pieces that are analyzed and that the appropriate rate for calculation of the total oxidation time is the intrinsic chemical rate reported by Los Alamos National Laboratory, which ranges from 1.0 to 2.1 kg Pu/m^2/min.[5,7,8,9]

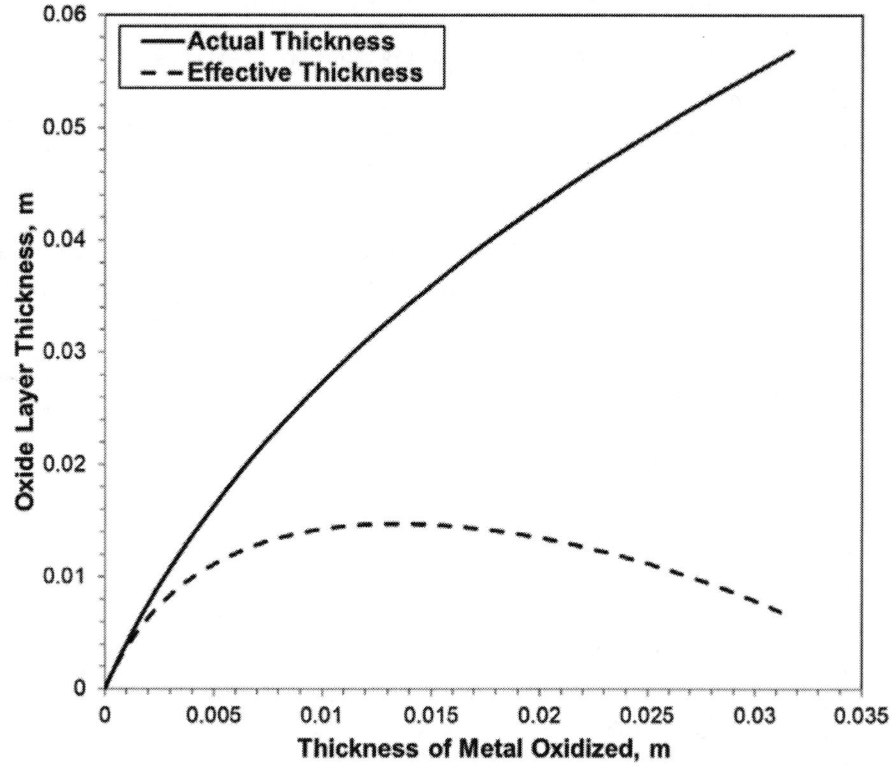

Figure 1. Variation of Oxide Layer Thickness during Oxidation of a 0.0889-m Diameter by 0.0318-m Thick Cylindrical Piece

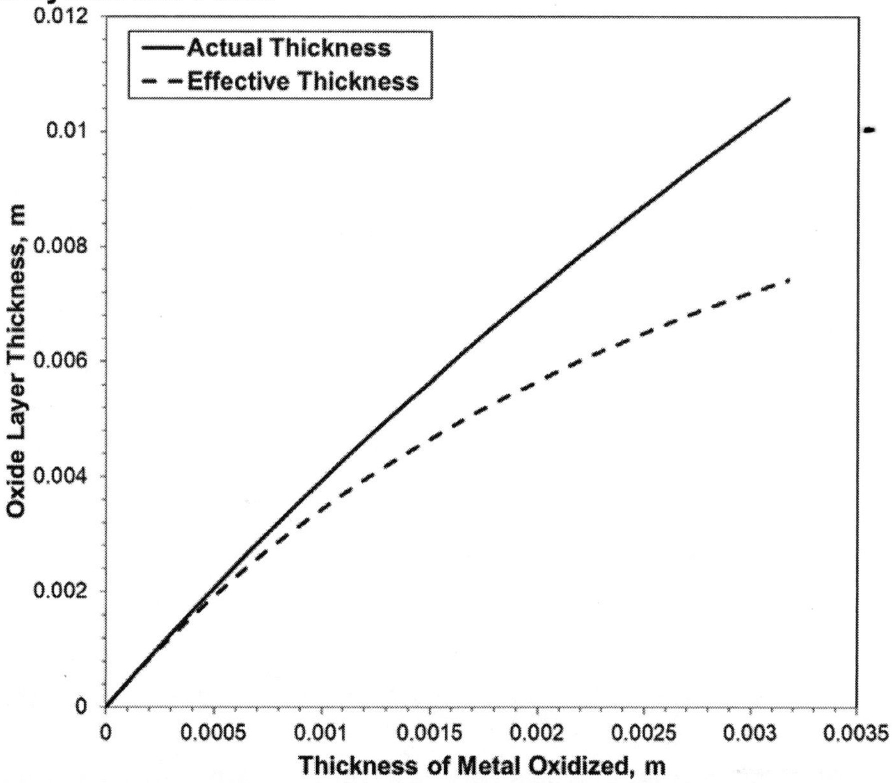

Figure 2. Variation of Oxide Layer Thickness during Oxidation of a 0.1016-m Square by 0.00318-m Thick Foil

Results of Oxidation Rate Calculations

Heat transfer calculations were performed to determine the limiting oxidation rates at varying maximum oxide temperatures inside a typical Savannah River Site muffle furnace, which has the dimensions of a 0.305-m cube, with 0.076-m thick alumina fiber walls. The heat transfer calculations assume that heat losses from the oxide to the furnace walls occur primarily by thermal radiation; the emissivity of the plutonium oxide was set at 0.9, and the emissivity of the alumina furnace walls was set at 0.42. The alumina fiber walls have a thermal conductivity of 0.10 W/m/K. A typical lower bound heat transfer coefficient of 5.0 $W/m^2/K$ was used to estimate external heat losses. Heat transfer to the air entering the furnace was modeled using natural convection correlations for enclosures.[10] The heat of combustion of the plutonium metal was set equal to the heat of formation for PuO_2, which is -1056 kJ/mol.[11]

The heat transfer calculations show that, for the average reported oxidation of 1.4 kg $Pu/m^2/min$, the furnace is too well insulated to allow for oxidation without triggering a temperature excursion. Figures 3 and 4 compare intrinsic rates of oxidation of the cylindrical "button" and the foil with the rates that can be achieved when the furnace is limited to 873 K and 1273 K. The results of the heat transfer calculations demonstrate that the oxidation reaction must be throttled by limiting the air flow into the furnace. The temperature-limited oxidation rate is about half the intrinsic rate at an oxide temperature of 1273 K and less than one-fifth of the intrinsic rate at an oxide temperature of 873 K.

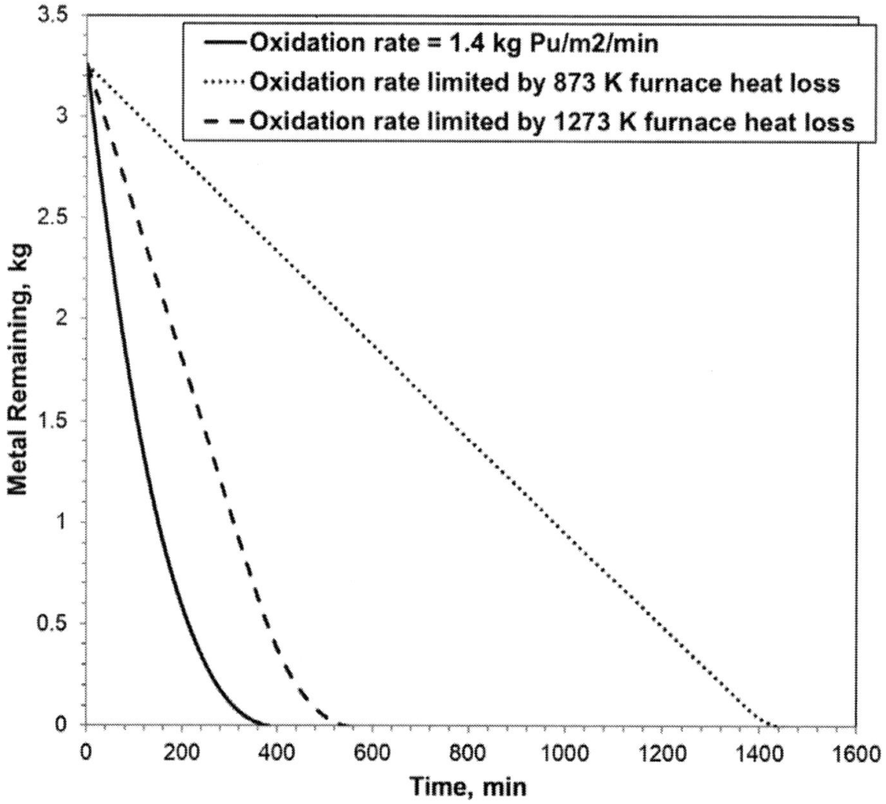

Figure 3. Comparison of Intrinsic Oxidation Rate for 0.0889-m Diameter by 0.0318-m Thick Cylindrical Piece with Rates Limited by Air Supply Rate

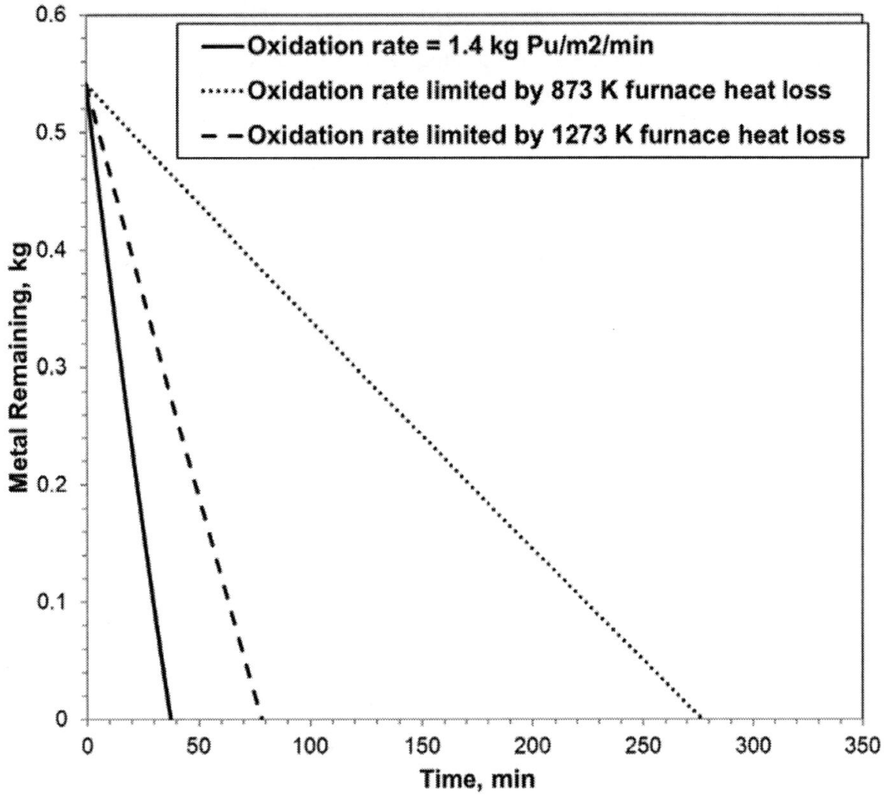

Figure 4. Comparison of Intrinsic Oxidation Rate for 0.1016-m Square by 0.00318-m Thick Foil with Rates Limited by Air Supply Rate

References

1. Katz, Joseph J., Glenn T. Seaborg, and Lester R. Morss, eds. (1986), The Chemistry of the Actinide Elements, 2nd ed., Vol. 1, London and New York, NY, Chapman and Hall, p. 610.

2. Musgrave, Larry E. (1971), "Effect of Water Vapor, Reduced Oxygen Concentrations, and Solvent Vapors on Plutonium Ignition," RFP Report RFP-1566, Dow Chemical Company, Rocky Flats Division, January.

3. Kolman, David G. (2002), "Direct Metal Oxidation Furnace System Description," LANL Report LA-UR-02-6627.

4. Haschke, John M., Thomas H. Allen, and L. A. Morales (2000), "Surface and Corrosion Chemistry of Plutonium," Los Alamos Science, 26.

5. Haschke, John M., and John C. Martz (1998), "Oxidation Kinetics of Plutonium in Air from 500 to 3500 °C, Application to Source Terms for Dispersal," J. Alloy Compd., 266 (1-2), pp. 81-89, February.

6. Bird, R. Byron, Warren E. Stewart, and Edwin N. Lightfoot (1960), Transport Phenomena, New York, NY, John Wiley & Sons, Inc., p. 200.

7. Haschke, John M., Thomas H. Allen, and John C. Martz (1998), "Oxidation Kinetics of Plutonium in Air: Consequences for Environmental Dispersal," J. Alloy Compd., 271-273, pp. 211-215, June.

8. Stewart, K. (1961), "Vixen A Trials: Experiments to Study the Release of Particulate Materials during the Combustion of Plutonium, Uranium, and Beryllium in a Petrol Fire," AWRE ReportTt15/60, Atomic Weapons Research Establishment, Aldermastion, UK.

9. Mishima, J. (1966), "Plutonium Release Studies: II. Release from Ignited, Bulk Metallic Pieces," Battelle Memorial Institute, Pacific Northwest Laboratories Report BNWL-357, November.

10. Bergman, Theodore L., Adrienne S. Lavine, Frank P. Incropera, and David P. DeWitt (2011), Fundamentals of Heat and Mass Transfer, 7th ed., Hoboken, NJ, John Wiley & Sons, Inc.

11. Katz, Joseph J., Glenn T. Seaborg, and Lester R. Morss, eds. (1986), The Chemistry of the Actinide Elements, 2nd ed., Vol. 2, London and New York, NY, Chapman and Hall, p. 1301.

529163 Study of Tritium Solubility and Diffusivity in Lithium Aluminate and Lithium Zirconate Pellets in Tpbar Using First Principle Density Functional Theory

Abstract:

Study of Tritium Solubility and Diffusivity in Lithium Aluminate and Lithium Zirconate Pellets in TPBAR Using First Principle Density Functional Theory

Hari P. Paudel, Yueh-Lin Lee, Yuhua Duan

National Energy Technology Laboratory, United States Department of Energy, Pittsburgh, Pennsylvania 15236, USA

Abstract

The ceramic materials γ-$LiAlO_2$ and Li_2ZrO_3 have superior thermo-physical and thermo-chemical properties, and are highly compatible with other blanket materials being used in the form of annular pellets in tritium-producing burnable absorber rods (TPBARs) to produce tritium (^3H or T) by thermal neutron irradiation of ^6Li. These materials have an excellent irradiation behavior at high temperatures, and are better swelling resistant than many other Li rich materials.[1-3] However, the transport mechanism of ^3H through the ceramic pellets and the barrier is hampered by the lack of data such as the diffusivity and solubility of hydrogen isotope. The study of Li diffusivity in Li containing ceramics, such as γ-$LiAlO_2$ and Li_2ZrO_3, has also been a subject of interest in Li-ion battery and solid oxide fuel cells. One of the powerful experimental techniques to understand atomic diffusion in crystals is nuclear magnetic resonance (NMR) measurement that probes the ion dynamics as a function of temperature. With the known value of spin-lattice relaxation rate from NMR measurements, it is possible to calculate the ionic jump rate (τ^{-1}) that allows us to further calculate diffusion coefficients using the Einstein−Smoluchowski equation.[4] In the past, by using NMR and dc conductivity measurements in γ- $LiAlO_2$, it was found that the Li diffusion coefficient is in the range of 10^{-20} to 10^{-13} m^2/s in temperature range of 400 K to 1000 K.[5] In the same experiment, Li activation energy was found to vary between 0.74 to 1.14 eV. Okuno and Kudo experimentally studied tritium diffusion and recovery processes in ceramic breeders irradiated with neutrons.[6] Their study revealed that the tritium produced in the ceramic materials (Li_2O, γ-$LiAlO_2$, Li_2SiO_3, Li_4SiO_4, Li_2ZrO_3, Li_8ZrO_6) irradiated with neutrons was released mainly in the chemical form of tritiated water (HTO) when heated in vacuum.

Despite several experimental and theoritical efforts aforementioned here on Li ion dynamics, roles of interaction between ^3H and vacancies on diffucivity, and its release behavior, and diffision mechanism of ^3H in pristine and radiation damaged crystal of $LiAlO_2$ and Li_2ZrO_3. After recoil of ^3H by losing several hundreds of eV in neclear reaction, ^3H remains as an impurity in the material. In addition to that, higly energetic alpha particles produced during the nuclear reaction process, sweep through the materials and can create vacancies and interstitial ions. These vacancies and interstitial ions can interact with a ^3H, resulting a significant change in ^3H diffusion and release behavior than expected. As a matter of fact, it is still an open question that which mechanism best describes the activation energy barrier of ^3H measured in the NMR experiments.[7]

In order to identify the mechanisms associated with atomic ^3H formation, diffusion, transport, deposition, and the kinetics in TPBAR pellets, in this study, we present the results of first-principles density functional theory (DFT) calculations of interstitial and substitutional ^3H defects, hydroxide (OT) vacancy defects, interaction of ^3H with O-vacancies in ceramics γ-$LiAlO_2$ and Li_2ZrO_3, and provide an understanding of how such defects hamper the diffusivity and solubility of ^3H.[7-8] This study focuses on the diffusion of ^3H and its species (such as OT^-, T_2, T_2O) in bulk γ-$LiAlO_2$ and Li_2ZrO_3 in presence of Li and O defects. In particular, we focus on the diffusion of substitutional and interstitial ^3H, its species (such as T_2O) and Li diffusion pathways in pristine and defective γ-$LiAlO_2$ and Li_2ZrO_3 crystal. We examine possible activation energy barrier of ^3H by computing several diffusion pathways. Our results show that the smallest activation energy barrier corresponds to the substitutional ^3H diffusion with barrier height of 0.63 eV in $LiAlO_2$ and 0.30 eV in Li_2ZrO_3 as shown in the following figures **a** and **b**. The reported experimental value of activation energy barrier for ^3H in $LiAlO_2$ is 0.93 eV.[6] The smallest OT diffusion barrier is found to be 2.17 eV in $LiAlO_2$ and that of 1.05 eV in Li_2ZrO_3 crystal. DFT results are more accurate at 0 K with experimental results obtained at low temperature. At elevated temperature, diffusion coefficient is expected to deviate due to contribution from the entropy term. However, temperature dependence of enthalpy and entropy is shown to be nearly cancelled out to each other.[9] In addition to that due to inherent limitation in correlation treatment with DFT, there could still be error while obtaining the diffusion barriers by subtracting transition image energy from the final state energy. Nevertheless, our results for diffusion barriers and diffusion coefficients in the ceramics γ-$LiAlO_2$ and Li_2ZrO_3 provide good estimates, and understanding of main diffusion mechanisms that exist in defectives and pristine pellets.

References:

1. Cao, H.; Xia, B. J.; Zhang, Y.; Xu, N. X., $LiAlO_2$-Coated $LiCoO_2$ as Cathode Material for Lithium Ion Batteries.

Solid State Ionics **2005**, *176*, 911-914.

2. Li, L. J.; Chen, Z. Y.; Zhang, Q. B.; Xu, M.; Zhou, X.; Zhu, H. L.; Zhang, K. L., A Hydrolysis-Hydrothermal Route for the Synthesis of Ultrathin $LiAlO_2$-Inlaid $LiNi_{0.5}Co_{0.2}Mn_{0.3}O_2$ as a High-Performance Cathode Material for Lithium Ion Batteries. *J. Mater. Chem. A* **2015**, *3*, 894-904.

3. Senor, D. J. *Recommendations for Science and Technology in Support of the Tritium Sustainment Program, PNNL-27216*; PNNL: 2017.

4. Shewmon, P., *Diffusion in Solids*; Springer International: Switzerland, 2016.

5. Indris, S.; Heitjans, P.; Uecker, R.; Roling, B., Li Ion Dynamics in a $LiAlO_2$ Single Crystal Studied by Li-7 Nmr Spectroscopy and Conductivity Measurements. *J. Phys. Chem. C* **2012**, *116*, 14243-14247.

6. Okuno, K.; Kudo, H., Tritium Diffusivity in Lithium-Based Ceramic Breeders Irradiated with Neutrons. *Fusion Eng. Des.* **1989**, *8*, 355-358.

7. Paudel, H.; Lee, Y.-L.; Holber, J.; Sorescu, D. C.; Duan, Y. *Fundamental Studies of Tritium Solubility and Diffusivity in $LiAlO_2$ and Lithium Zirconates Pellets Used in TPBAR*; DOE-National Energy Technology Laboratory, DOE/NETL-PUB-21464: 2017.

8. Paudel, H.; Lee, Y.-L.; Senor, D. J.; Duan, Y., Tritium Diffusion Pathways in γ-$LiAlO_2$ Pellets Used in TPBAR: A First-Principles Density Functional Theory Investigation. *J Phys Chem C* **2018**, accepted.

9. Mantina, M.; Wang, Y.; Arroyave, R.; Chen, L. Q.; Liu, Z. K.; Wolverton, C., First-Principles Calculation of Self-Diffusion Coefficients. *Phys. Rev. Lett.* **2008**, *100*, 215901.

529818 Analysis of Helium Segregation on Surfaces of Plasma-Exposed Tungsten

Abstract:

Plasma facing components (PFCs) as the first wall materials in nuclear fusion reactors are exposed to intense plasma heat and particle fluxes. The implantation of helium (He) atoms into these materials impacts significantly their surface morphological and near-surface structural evolution. Tungsten (W) is a promising PFC material because of its thermomechanical properties. In tungsten, such interstitial He atoms are very mobile and aggregate to form clusters of various sizes. Small, mobile helium clusters (He_n, 1â‰¤*n*â‰¤7) are attracted to the tungsten surface due to an elastic interaction force that provides the thermodynamic driving force for surface segregation and their diffusional transport mediates the evolution of surface morphology and near-surface microstructure.

In this presentation, using molecular-statics (MS) computations and molecular-dynamics (MD) simulations based on a reliable many-body interatomic potential, we explore helium segregation on tungsten surfaces for tungsten that has been exposed to different levels of helium implantation. The cluster-surface elastic interaction force induces drift fluxes of these mobile He_n clusters, which increase substantially as the migrating clusters approach the surface, facilitating helium segregation on the surface. Moreover, the clustersâ€™ drift toward the surface enables cluster reactions in the near-surface region at rates much higher than those in the bulk material; the most important of these reactions is trap mutation, which produces W surface adatoms and immobile complexes of helium clusters surrounding W vacancies located within the lattice planes at a short distance from the surface that evolve to form He bubbles. With higher levels of helium implantation at higher fluence, these mobile clusters are subjected to cluster-cluster and cluster-bubble interactions in addition to the more dominant cluster-surface interactions, which complicate cluster dynamics beyond the dilute limit of helium content in the PFC material. These near-surface cluster dynamics have significant effects on the surface morphology, near-surface defect structures, and the amount of helium retained in the material upon plasma exposure.

We characterize in detail helium cluster dynamics and their effects on helium surface segregation based on analysis of a large number of MD trajectories for such mobile clusters near W(100), W(110), W(111), and W(211) surfaces. Furthermore, we analyze the effects of varying helium fluence due to increased plasma exposure of the tungsten PFC by performing systematic MS computations of small helium cluster energetics near low-Miller-index W surfaces as a function of distance (depth) of the cluster center from the surface on a grid of lateral locations on the surface. We analyze the defect interactions that mediate the energetics of small He clusters migrating to the surface away from the dilute limit of He in W taking into account that the migrating cluster also is subjected to the stress field generated by larger He bubbles, as well as other He clusters, and quantify the strength of these interactions for different levels of He implantation. The outcome of this analysis is the systematic parameterization of helium mobile cluster energetics at varying levels of He implantation through functional forms that include contributions from cluster-cluster and cluster-bubble interactions as well as cluster-surface interactions. Such parameterizations are important for developing atomistically informed, hierarchical multiscale models of helium cluster dynamics in PFC materials.

527775 Nonlinear Dynamics of Bubble Collapse

Abstract:

Nonlinear Dynamics of Bubble Collapse

Jyoti Bhati[1] , Swapan Paruya[1], S. Pushpavanam[2]

National Institute of Technology Durgapur, West Bengal, India -713209

Indian Institute of Technology Madras , Chennai,Tamil Nadu, India -600036

Extended Abstract

Boiling has many applications in the field of nuclear energy, refrigeration, microelectronic and aerospace. In the boiling, study of bubble dynamics is very important due to its complexity and its direct effect on the design of heat transfer equipment. Bubble collapse is an important and very complex phenomenon of the bubble dynamics. Bubble collapse occurs only in subcooling boiling, in which bubbles are surrounded by subcooled water. The degree of subcooling significantly affects the rate of bubble collapse .When a bubble collapses, it release high thermal energy from bubble to liquid. This particular aspect of bubble collapse is also utilized in cavitation reactors, which requires a very temperature. However, this energy transfer during bubble collapse creates a number of instability phenomena in the boiling channel and affects the interfacial heat transfer coefficient. It is noted that literature lacks steady state analysis of oscillating bubble collapse. In this paper, the authors report the steady-state analysis and the bifurcation analysis of bubble collapse at a pressure of 1.0 atm.

Model Equations

We have used Plesset-Zwick equation (Eq. (1)) and the energy balance equation (Eq. (3)) at bubble wall for the steady sate analysis. Plesset-Zwick equation (modified Rayleigh-Plesset equation) describes the growth or collapse of a spherical vapor-bubble. $\frac{dR}{dt} = y$ (eq. (2)) is the growth rate of bubble that produces liquid inertia force. $P_v = P_{sat}(T_\infty)$ is the vapor pressure at liquid temperature (T_∞); P_∞ is the ambient pressure; σ is the surface tension.; h is interfacial heat transfer coefficient.

$$\frac{P_v(T_v)-P_\infty}{\rho_l} = R\frac{dy}{dt} + \frac{3}{2}(y)^2 + \frac{2\sigma}{\rho_l R} \quad (1)$$

$$\frac{dR}{dt} = y \quad (2)$$

$$\frac{d\left(\frac{4}{3}\pi R^3 \rho_v h_{fg}\right)}{dt} = -4\pi R^2 (T_v - T_\infty) \quad (3)$$

The energy required for evaporating the liquid is supplied by thermal diffusion through the liquid. For the spherical bubble, energy balance for moving liquid at vapor-liquid interface is given by Eq. (3):.

We linearized the above three equation to obtain the following form:

$$\frac{dx}{dt} = Ax + B \quad (4)$$

where, $x = [R, dR/dt, T_v]^T$

We have performed the steady state analysis by arc-length continuation method. And we also determined the eigenvalues of Jacobian matrix (A) for studying bifurcation phenomena during bubble collapse. Using bifurcation theory, we study how a solution branch breaks up into two, giving rise to another class of solutions, as we vary a parameter. The steady sate of a system is determined by a large number of parameter which occur in the model. The dependency of a steady sate on a parameter is generated by varying it and keeping all other constant. This enables us to study how a steady state solution change. We have used arc – length continuation method to trace the steady state branches. We have determined set of eigenvalues on the steady-state branches. From the eigenvalue, we can clearly understand the stability or oscillating nature of bubble collapse.

Numerical Results

In the present study on the stability analysis of the bubble collapse, we have T_∞ (subcooled water temperature) and R (initial bubble radius) to be the bifurcation parameters. We obtained the variation of T_v (vapour temperature) vs. R (by arc -length continuation method) and eigenvalues for every T_v and R. From the results presented in Fig.1, we see that the vapour temperature increases as R is increased if interfacial heat transfer coefficient is constant (h=1000W/m^2-K). In Fig .2 we see that the vapour temperature decreases as initial bubble radius is increased, if interfacial heat transfer coefficient varies with R (Nu=hD/k_l=2 for spherical bubble collapse). The eigenvalues indicate that unstable bubble oscillations take place during bubble collapse. Smaller bubbles create relatively more unstable phenomena during collapse. As initial bubble radius is increased, we obtain unstable nodes – the unstable decrease of radius without oscillation.

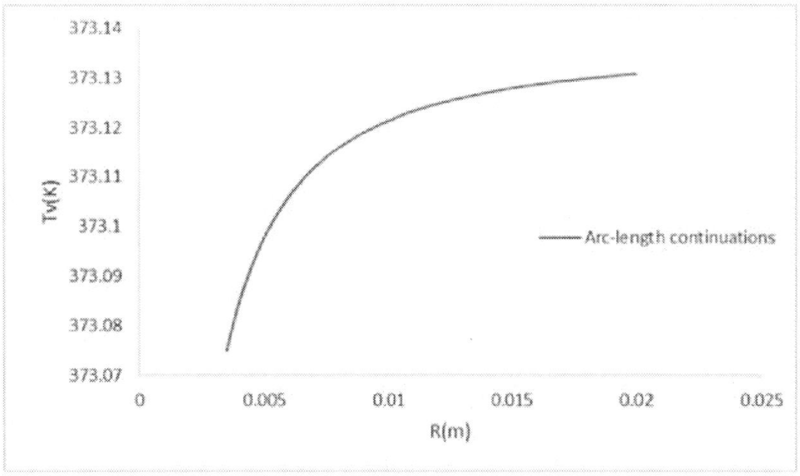

Fig 1. Steady -state of T_v and R by Arc length continuations method where h=1000W/ (m^{2-}K) at $\Delta T_{sub} = 12.2_0$K

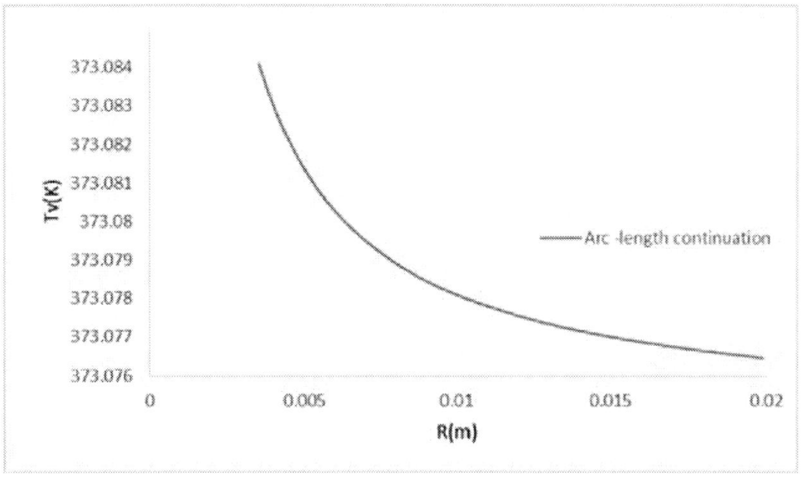

Fig.2 Steady –state analysis of bubble collapse by using arc-length continuation method where (Nu=2) at $\Delta T_{sub} = 12.20_K$

Keywords: Bubble, Bubble collapse, Stability analysis, Bifurcations, Arc- length continuation

531398 Density Functional Theory Study of the Tritium Formations on the Surfaces of γ-$LiAlO_2$

Abstract:

One of the important components of tritium-producing burnable absorber rods (TPBAR) is the γ-$LiAlO_2$ layer used in the form of an annular ceramic pellet enriched with the ^6Li isotope. Within TPBAR architecture, this oxide is located between the zircaloy-4 liner and nickel-plated zircaloy-4 tritium getter. When irradiated in a pressurized water reactor (PWR), the ^6Li pellets absorb neutrons, simulating the nuclear characteristics of a burnable absorber rod, and produce tritium (T). The T specie chemically reacts with the metal getter, leading to the formation of a metal hydride. Accurate analysis of the T transport through the ceramic pellets and the barrier/cladding system is hampered by the lack of fundamental data about the hydrogen isotope solubility and diffusivity. In TPBAR architecture, between the γ-$LiAlO_2$ pellet and the Ni-plated Zircaloy-4 getter (or liner), there is a gas gap. After T diffuses from bulk γ-$LiAlO_2$ to its surface, it is possible that different T species (such as T, T_2, T_2O, and OT, etc.) are formed on this surface which can then get into the gas gap region. Furthermore, upon recombination and desorption the molecular tritium can be formed on the γ-$LiAlO_2$ surface. Therefore, exploring the formation of T-related species on γ-LiAlO2 surface becomes very important as an integral part of the study of T solubility and diffusivity in the pellets. Based on the density functional theory, in this study, we investigate the stability and structural properties of γ-$LiAlO_2$ surface, the most possible adsorption sites, and the formation of T on surface. From the calculated surface free energies, the most stable surfaces were identified from all possible symmetrical stoichiometric and nonstoichiometric low-index surfaces. By calculating the binding energies of T adsorbed on top, bridge, and hollow sites of $LiAlO_2$ (100) surface, we obtained four stable adsorption sites (three O-top sites and one Al-top site) and the most stable one is O-top site. From the Bader charge analysis, T adsorption on O-top site is a reduction process, while on Al-top site is an oxidation one. In addition, the zero-point energy correction (ΔE_Z) on the stability of adsorption is investigated. The obtained results showed that the ΔE_Z influence of T adsorption on O-top site is larger than that on Al-top site, but the trend of stability keeps unchanged.

530785 Modeling of Surface Morphological Evolution of Plasma-Facing Tungsten in Fusion Reactors

Abstract:

Tungsten (W) and tungsten alloys are widely viewed as the most promising plasma facing material (PFM) candidates for divertor and first-wall systems in a nuclear fusion reactor. However, experiments aiming to investigate PFM performance under continuous operation and high first-wall temperature have shown that a nanostructure with a fuzz-like morphology develops on the W surface under the operating conditions of temperature, helium (He) impact energy, and He flux expected for ITERâ€™s divertor. Formation of such fuzz nanostructure may adversely influence the reactor performance and operation. The aim of our work is to obtain a fundamental understanding of the initial stage of fuzz formation by modeling the surface morphological evolution of helium-ion-irradiated tungsten considered as a PFM.

We have developed an atomistically-informed (with constitutive equations parameterized using results of large-scale molecular-dynamics simulations [1]), continuous-domain model to describe the surface morphological evolution of helium-ion-irradiated tungsten. Based on this model, we have conducted self-consistent numerical simulations of the dynamics of the irradiated W surface morphology and benchmarked the simulation results by systematic comparisons with experimental measurements. Under the experimental conditions, namely, an RF plasma source ($\sim 3\times 10^{20} m^{-2} s^{-1}$) exposure of 75-eV He on ITER-grade W at 840°C, the surface morphology predicted by our simulations is in good qualitative agreement with the experimental observations for the early stage of nanotendril formation on the W surface, a precursor to fuzz-like surface growth. Additionally, quantitative comparisons between the experimental data and the simulation results show that our model predicts the growth rate of the nanotendrils reasonably well, and provides reasonable approximate estimates for the nanotendril width and arrangement on the W surface. Two plausible factors affecting the predictive capabilities of our model are hypothesized: subsurface bubble dynamics and bubble bursting/pinhole formation. Future model extensions to test these hypotheses, as well as the anticipated divertor performance are discussed. Furthermore, we have investigated the dependence of the nanotendril width and nanotendril arrangement on the W surface on the He fluence and the W surface temperature.

[1] K. D. Hammond, S. Blondel, L. Hu, D. Maroudas, and B. D. Wirth, *Acta Materialia* **144**, 561-578 (2018).

531738 Stability and Interactions of Point Defects in Lithium Metal Oxides for the Tritium-Producing Burnable Absorber Rod Applications

Abstract:

Point Defect Energetics in Lithium Metal Oxides for Tritium-Producing Burnable Absorber Rod Applications

Yueh-Lin Lee[1], Jamie Holber[1], Hari P. Paudel[1], Dan C. Sorescu[1], and Yuhua Duan[1]

[1]*National Energy Technology Laboratory, 626 Cochrans Mill Road, P.O. Box 10940, Pittsburgh, Pennsylvania 15236-0940, USA*

Ceramics, such as γ-$LiAlO_2$, Li_2TiO_3, and Li_2ZrO_3, are candidate systems for their suitability as blanket materials in tritium-producing burnable absorber rods (TPBARs). After adsorbing neutrons, ^6Li is readily converted to tritium through reaction [1]. The oxide defects and structural disorder may enhance the diffusion of Li and tritium species within TPBAR materials. Previous experimental results have indicated that interactions between tritium and irradiation-induced defects could play a key role in the observed tritium solubility and diffusivity and overall can modify the tritium release from TPBAR materials [2,3]. Furthermore, previous theoretical studies in bulk Li_2O and Li_2TiO_3 [4-6] have suggested that both Li and O vacancies can act as trapping sites for tritium species.

In this work, the phase stability and defects energetics of $Li_xM_yO_z$ (M=Al, Ti, and Zr) materials are investigated using density functional theory (DFT) calculations and *ab initio* thermodynamic analysis. Theoretical investigations were performed for several types of charged defects including vacancies, interstitials, antisite defects, and ^3H substitutional defects to identify the trends of the point defects stability under TPBAR operating temperature and oxygen partial pressure conditions [7,8]. The interplay between defects stability, defects interactions, the change of the environmental chemical potentials, and the intrinsic Fermi energy level under the TPBAR operating conditions are discussed. As shown in the Figure 1 below, our results suggest that interactions between the tritium and the point defects are dependent on the electron chemical potential across the band gap, *i.e.* the Fermi energy level. By considering the intrinsic Fermi level position for a range of external conditions, *i.e.* $P(O_2)=10^{-5}$ atm at T=1000K to $P(O_2)=10^{-25}$ atm at T=1000K, the results obtained suggest that γ-$LiAlO_2$ phase exhibits stronger interactions between the ^3H interstitial and the V_{Li}, V_O, and O_{int} defects than Li_2TiO_3 and Li_2ZrO_3 materials, indicating a higher tendency to trap ^3H at these point defects [7] and correspondingly a lower efficiency for tritium recovery in TPBARs.

Among the three TPBAR materials investigated, *i.e.*, Li_2TiO_3, Li_2ZrO_3, and γ-$LiAlO_2$, our results indicate that the point defect formation energies are higher in γ-$LiAlO_2$ than in the other two materials. The increased defect stability in γ-$LiAlO_2$ correlates also with the experimental tritium/hydrogen diffusivities which are the slowest in this material. At the same time, the higher point defect formation energies observed in γ-$LiAlO_2$ indicate a greater stability of this material with the possibility to better withstand the long-term irradiation at high temperatures in a reactor. Our theoretical results provide thermodynamic guidance for the factors governing the tritium transport properties [9], point defect energetics and the phase stability in $Li_xM_yO_z$ (M=Al, Ti, and Zr) systems under TPBAR operating chemical potential conditions [10].

References:

[1] Senor, D.J., *Recommendations for Science and Technology in Support of the Tritium Sustainment Program, PNNL-27216*. 2017, PNNL

[2] Moriyama, H.; Tanaka, S.; Noda, K., *Irradiation effects in ceramic breeder materials*. Journal of Nuclear Materials, 1998. **258**(Part 1): p. 587-594.

[3] Kobayashi, M., et al., *Dependency of irradiation damage density on tritium migration behaviors in Li2TiO3*. Journal of Nuclear Materials, 2014. **447**(1): p. 1-8.

[4] Shah, R.; Devita, A.; Payne, M.C., *Ab-Initio Study of Tritium Defects in Lithium-Oxide*. Journal of Physics-Condensed Matter, 1995. **7**(35): p. 6981-6992.

[5] Murphy, S.T., *Tritium Solubility in Li_2TiO_3 from First-Principles Simulations*. The Journal of Physical Chemistry C, 2014. **118**(51): p. 29525-29532.

[6] Tanigawa, H.; Tanaka, S., *Ab-initio study on interaction of hydrogen isotopes with charged defects in lithium oxide*. Journal of Nuclear Materials, 2002. **307-311**: p. 1446-1450.

[7] Lee, Y.-L; Holber, J.; Paudel, H. P..; Sorescu, D. C.; Sensor, D.; Duan, Y., "*Density Functional Theory Study of the Point Defect Properties for the γ-$LiAlO_2$, Li_2ZrO_3, and Li_2TiO_3 Materials*", (2018) to be submitted.

[8] Broberg, D., et al., PyCDT: A Python toolkit for modeling point defects in semiconductors and insulators. Computer Physics Communications, 2018. **226**, p. 165-179

[9] Paudel, H. P., Lee, Y. -L., Senor, D. L., Duan, Y., "*Tritium Diffusion Pathways in γ-$LiAlO_2$ Pellets Used in TPBAR: a first-principles density functional theory investigation*", Journal of Physical Chemistry C (2018) revision submitted.

[10] Paudel, H. P., Lee, Y.-L., Holber, J., Sorescu, D. C., Duan, Y., "*Fundamental Studies of Tritium Solubility and Diffusivity in $LiAlO_2$ and Lithium Zirconates Pellets Used in TPBAR*", Tritium Science Program FY17 Report, DOE/NETL-PUB-21464, Nov. 2017

536813 Effects of Flux on Helium Bubble Growth in Plasma-Facing Materials

Abstract:

Tungsten, the primary material under consideration as the divertor material in magnetic-confinement nuclear fusion reactors, has been known for the last decade to form "fuzz"—a layer of microscopic, high-void-fraction features on the surface—after only a few hours of exposure to helium plasma. Typical molecular dynamics simulations of phenomena relevant to plasma-facing tungsten are based on fluxes of $\sim10^{28}$ m^{-2} s^{-1}, while fluxes in experiments are typically closer to 10^{21}–10^{23} m^{-2} s^{-1}. Large-scale molecular dynamics simulations of post-implantation helium behavior in plasma-facing tungsten single crystals recently revealed orientation-dependent depth profiles, surface evolution patterns, and other crystallographic and diffusion-related characteristics of helium behavior in tungsten during the first microsecond. This study tackles the issue of the effect of helium flux by way of long-time (500–1000 ns) molecular dynamics simulations of plasma-facing tungsten at fluxes $\sim10^{27}$, 10^{26}, and 10^{25} m^{-2} s^{-1}, all whilst tracking surface features and helium depth profiles. These calculations serve as important benchmarks for coarse-grained simulations and unveil the relative importance of the transport processes and material deformation process involved in surface evolution in plasma-facing materials.

533107 The Long-Term Evolution of H/He Irradiated in W By a Multi-Scale Approach

Abstract:

A sequential multi-scale approach combining Monte Carlo, Density functional theory/Molecular dynamics and Cluster dynamics (MC-DFT/MD-CD) has been developed to effectively couple atomic diffusion events with displacive and continuous processes at finite temperature, which enables us to investigate the long term and large scale behavior of the irradiated ions in nuclear device related materials. The evolution of different types of defects with time and depth is obtained in comparison with experiments. Specifically, the cases of H and He in tungsten are discussed in details, where the evolution of different types of defects with time and depth is investigated. The contribution to the ion concentration and the depth profile could provide the understanding on the dominated factors among the multi-micromechanisms such as trapping, diffusion, grain boundary and dislocation, etc., which is obviously important for estimating damages to the nuclear materials.

Referencesï¼š

1. Y. G. Li, Y. Yang, M. P. Short, Z. J. Ding, Z. Zeng and J. Li, *Sci. Rep.* 5 (2015) 18130.
2. http://theory.issp.ac.cn/IM3D.
3. Y. G. Li, W. H. Zhou, R. H. Ning, L. F. Huang, Z. Zeng and X. Ju, *Commu. Comput. Phys.* 11 (2012) 1547.
4. C. G. Zhang, W. H. Zhou, Y. G. Li, Z. Zeng, X. Ju, *J. Nucl. Mater.* 458 (2015) 138.
5. Y. G. Li, W. H. Zhou, L. F. Huang, Z. Zeng and X. Ju, *J. Nucl. Mater.* 431 (2012) 26.
6. Z. Zhao, Y.G. Li, C.G. Zhang, G.Y. Pan, P.F. Tang and Z. Zeng, *Nucl. Fusion* 57 (2017) 086020.
7. R. H. Ning, Y. G. Li, W. H. Zhou, Z. Zeng, and X. Ju, *J. Nucl. Mater.* 430 (2012) 20.
8. L. Hu, Y.G. Li, C.G. Zhang and Z. Zeng, *RSC Adv.* 5 (2015) 65750.
9. W. H. Zhou, C. G. Zhang, Y. G. Li, and Zhi Zeng, Sci. Rep. 4(2014).

538608 Design and Implementation of a Nuclear Solvent Extraction Plant-Level Simulator

Abstract:

There exists interest in the development of robust, dynamic, plant-level solvent extraction simulators that can incorporate multi-phase, multi-radionuclide species, rate-base reactive processes in a flowsheet configuration. This presentation describes progress towards developing such a simulator for use with realistic aqueous/organic phases typical of nuclear reprocessing and/or radio-isotope production. The model is based on a dynamic network approach which decouples individual contactor elements in a countercurrent cascade of contactors. The time-integration of the DAE's proceed by synchronizing nodes in the network periodically. This approach facilitates the implementation of flowsheets that use bank of contactors such as centrifugal and/or mixer-settlers.

The rate-based modeling method allows for incorporation of non-equilibrium processes through mass transfer parameters that could be obtained via canonical experimental calibration in the future. We have been developing a software for solvent extraction that can be used as a plug-and-play simulator to investigate a more comprehensive dynamic behavior of plant facilities and/or laboratory scale.

523633 Development of an Infrared Measurement Method for Molten Fluoride Salt

Abstract:

Molten fluoride salts have been proposed as both coolant and/or fuel solvent in advanced nuclear reactor designs. In order for these designs to be commercially viable, chemistry control systems must be developed, due to the extremely corrosive behavior and high melting temperature of these salts. Critical to these systems will be the ability to detect and calculate activities of melt species. In principle, an infrared spectrometer could perform this task, through correlation of peaks with vibrational modes of various complex species. In addition, the high temperatures and large thermal gradients that are present in some designs could lead to situations where radiative heat transfer (RHT) would be a significant heat transfer mechanism. In order to model RHT accurately, precise infrared measurements must be made to determine the spectral absorption coefficient of the melt of interest. We are working to develop emission and reflection infrared measurement techniques, using a commercial high-temperature cell and reflection accessory. Preliminary data has been obtained and validated for solid reference materials, including fused SiO_2 (silica) and crystalline Al_2O_3 (sapphire). A preliminary design has been generated of a custom sample holder which will fit in the cell to minimize both the curvature of the salt surface and the degree to which the detector sees radiation from the sample container walls. Future work will include low-temperature measurements of solid LiF and FLiNaK, development of a gas phase control system, and high-temperature measurements of molten FLiNaK.

528116 Density Measurement of Molten Fluoride Salts

Abstract:

The measurement of molten fluoride salt density is important for the design of Fluoride Salt - Cooled High Temperature Reactors (FHRs). Density plays a crucial role in neutronics analysis, and it is essential to consider the temperature dependence of density in natural convection analysis for FHRs. Density measurements of two molten fluoride salt mixtures were performed at the University of Wisconsin (UW)-Madison: LiF-BeF_2 (FLiBe) and LiF-NaF-KF (FLiNaK) at temperatures near melting (459 ÌŠ and 454 ÌŠ respectively) to 800 ÌŠ C. Density measurement experiments took place inside an Argon glovebox in order to protect laboratory personnel from beryllium hazards and to maintain water and oxygen levels below 4 ppm to avoid salt contamination. A Mettler Toledo analytical balance XS104 was used for measurements. The scale comes equipped with a small hook underneath the top loading tray from which a wire is hanged. The opposite end of the wire is hooked to a nickel bobber which is then submerged in the molten salt. By using Archimedes principle and the known volumetric expansion of the nickel bobber, the density of the molten salt mixture is obtained from the buoyant weight measurement shown on the scale. Extensive calibrations for the experiment apparatus were performed for temperature and mass readings to reduce the amount of measurement uncertainty. The experimental data was then compared to reported literature values, as well as to a prediction of density using additive molar volumes of the constituents. Preliminary data collected for FLiBe shows agreement within 6% of density values from NIST measurements with a 0.2% uncertainty. Results for FLiNaK failed to yield a linear fit at temperatures near melting, from 456 to 525 ÌŠ C, which contributed to an uncertainty of 2.7%. Work is currently underway to produce repeatable results for FLiBe measurements and to remeasure FLiNaK density. Changes in the experimental apparatus and methodology will involve a verification of isothermal conditions in the salt sample during measurement and characterization of samples pre and post measurement.

539973 Design and Operation of Batch Fluoride Salt Purification Systems

Abstract:

Fluoride salts were extensively investigated at Oak Ridge National Laboratory (ORNL) for molten salt reactor (MSR) applications during the 1950â€™s-1970â€™s. In these reactor concepts, the fuel is contained the molten salt. Fluoride salts without fuel can also be used as coolants. Recently, there has been a resurgence of interest in MSRs. ORNL has revived and initiated several efforts in molten salt to support the research and development needed to advance MSR technology. Fundamental to these efforts are the ability to produce clean and characterized salt for subsequent research.

Molten fluoride salts are fluxing agents, in that they remove protective oxides scales. To inhibit corrosion, the salt needs to be purified to remove oxygen and moisture. Two facilities have been developed and brought online at ORNL to purify fluoride salts. Both systems are batch processes that utilize HF, H_2, and Ar as the process gases. The large system has been used to clean 160 kg batches of LiF-NaF-KF (FLiNaK) salt within a large nickel crucible. A smaller system has been used to clean a 3.2 kg batch of LiF-BeF2 (FLiBe) salt inside of a graphite crucible. This paper describes the design of these two purification systems and experiences and results from their initial operations.

537606 Modeling Radiative Heat Transfer in High-Temperature Liquid-Salts

Abstract:

INTRODUCTION

Recent technological advancements and needs for high-temperature heat have increased interest in both nuclear and solar high-temperature liquid-salt systems. Growing nuclear concepts include the Fluoride-salt-cooled High-temperature Reactor (FHR) that uses solid fuel and a liquid salt coolant, the Molten Salt Reactor (MSR) with fuel dissolved into the salt coolant, and high-magnetic-field fusion machines with immersion salt coolant blankets [1]. In parallel high-temperature concentrated solar power systems propose using liquid-salts as volumetric absorbers and for energy storage [2]. What these design have in common is that they utilize high-temperature chloride and fluoride salts that may experience the effects of participating media radiative heat transfer. While some work has been conducted on the thermophysical properties of these fluids [3], very little is known about their radiative properties. Because these salts operate between 600-1000C and radiative heat transfer increases as T^4, thermal radiation has the potential to become a significant heat transfer mechanism by decreasing surface structure temperatures while increasing temperature uniformity, especially in off-normal higher-temperature conditions. Therefore, new, innovative research in this area is required.

DESCRIPTION OF WORK

The purpose of this work is to create a methodology for understanding and modeling the effects of radiative heat transfer in high-temperature liquid-salt systems. First, spectral absorption coefficients of candidate salts will be measured using Fourier Transform Infrared Spectroscopy. Then, using that data, computational fluid dynamics (CFD) simulations, validated by scaled experiments, will be run quantify the impact of thermal radiation to the overall heat transfer. By demonstrating that CFD can capture the correct physics, complex geometries with new salts can be analyzed for use in various heat transfer applications.

BACKGROUND

At the temperatures of the above energy applications, liquid-salts may act as participating media where thermal radiation can be absorbed and re-emitted within the fluid. This heat transfer mechanism is controlled mainly by the absorption coefficient, k. Therefore, to model radiative heat transfer, k must be well characterized in regions of significant emissive power. For high-temperature applications, radiation is emitted by the salt itself between 600-1000C, so our focus will be on the near to mid IR wavelengths (2-12μm). Theoretical calculations have shown that some chloride and fluoride liquid-salt absorption edges may occur in that region [4]. Additionally, nuclear salts can contain impurities that can greatly increase the absorptivity of the salt in the near IR.

EXPERIMENTAL METHODS

1. Measurement of Absorption Coefficient

The spectral absorption coefficients, $k(\lambda)$, of candidate chloride and fluoride salts will be measured using Fourier Transform Infrared (FTIR) spectroscopy. The FTIR apparatus is composed of two, concentric nickel cuvettes with transparent, diamond windows, a light source, and light detector. Liquid-salt is placed in the outer cuvette while the inner cuvette is free to move up and down, allowing for variable attenuation thickness. A broad-spectrum light source is then passed through the diamond windows and attenuated by the liquid-salt. Using Beer's law and taking the ratio of attenuated intensities (I) at multiple thicknesses (Δx), all background attenuation can be cancelled out and the absorption coefficient of the salt can be isolated.

2. CFD Modeling with Experiments

CFD simulations are being conducted and will be validated by experimental testing to prove understanding and predictability of the heat transfer behavior. Historically, CFD radiation work has focused on surface-to-surface radiation with a small number of participating media scenarios (i.e. combustion gases) where the index of refraction is close to 1. This work will expand upon those studies and increase the applicability of CFD modeling to advanced liquid-salt concepts with participating media. This leads to enhancements in the accuracy of wall temperature predictions as well as calculated energy distributions within the fluid.

In this work, high-fidelity 3D CFD simulations are conducted using STAR-CCM+, a vetted commercial code. These simulations are run at high temperatures with participating media multiband thermal radiation models that have been validated against participating media coupled with conduction and convection test cases with known numerical solutions. Using STAR-CCM+, radiation models can be turned on and off to isolate the effects of radiation. These results are then confirmed by running experiments with hot, liquid-salts. However, because radiative heat transfer cannot be removed in real, high-temperature experiments, Dowtherm A, a well documented simulant fluid [5], is used to match important dimensionless parameters at lower temperatures. High-temperature liquid-salt and Dowtherm A experiments can then be directly compared to CFD simulations with and without thermal radiation.

Preliminary CFD work is currently being used to inform experimental design choices. By using this iterative CFD/experimental process, we are able to minimize potential sources of error as well as ensure experiments capture heat transfer and flow regimes of interest. Initial liquid-salt tank simulations with an immersion heater confirm that, by accounting for participating media, reduced surface temperatures and increased fluid temperature uniformity are observed.

PATH FORWARD

The apparatus designed to measure absorption coefficients of chloride and fluoride salts is currently under construction. Once results are obtained, they can then be used in CFD simulations to obtain preliminary results. Currently, theoretically calculated absorption coefficients are being used.

REFERENCES

1. FORSBERG, D. WHYTE, "Converging Fission and Fusion Systems Toward High-Temperature Liquid-Salt Coolants: Implications for Research and Development and Strategies," *Trans, ICCAP*, San Francisco, CA (2016).
2. SLOCUM ET AL., "Concentrated Solar Power on Demand," *Solar Energy* 85:1519-1529 (2011).
3. FORSBERG ET AL., "Liquid Salt Applications and Molten Salt Reactors," Revue Generale Nucleaire, 4 (2007).
4. CHALEFF ET AL., "Radiation Heat Transfer in Molten Salt FLiNaK," Nuclear Technology, 196 (2016).
5. THERMAL HYDRAULICS WORKING GROUP, "Fluoride-Salt-Cooled High-Temperature Reactor Benchmarking White Paper," University of California, Berkeley, (2016).

531254 Viability of Molten Salt Reactors for the Production of Molybdenum-99

Abstract:

This study examines the feasibility of producing molybdenum-99 in molten salt reactors (MSR). With the aging of current isotope-producing reactors and the rising demand for medical isotopes, future shortages are predicted to greatly increase the cost of Mo-99. Molten salt reactors have the potential for safe, efficient operation at a variety of scales. The liquid nature of the fuel makes processing the coolant salt without shutting down the reactor possible, a possibility which we will refer to as â€œ*in situ* processing.â€ Such processing could be used to reduce the waste vector of the fuel, including long-lived radioisotopes. It can also be used to remove valuable fission products, including medical isotopes such as Mo-99. However, the viability of the MSR for Mo-99 production has not been rigorously analyzed. While Mo-99 is a major fission product, it is unclear how accessible it will be for direct extraction from a fuel salt given its relatively short half-life.

In this study, a computationally inexpensive mathematical model is used to analyze fission product accumulation in an MSR, with the specific goal of assessing the feasibility of *in situ* Mo-99 extraction. The model includes generation and removal of all isotopes in an MSR from fission, decay, activation and *in situ* processing or separation. The nuclear interaction data used in the model are taken from the ENDF/B-VII.1 database. The reactor model is based on a homogenous (liquid-fueled) mono-energetic neutron spectrum core design. The influence of several important MSR operating parameters were examined including uranium concentration, neutron energy, and power density, in addition to parameters that describe the separation process. This separation was assumed to be mass-transfer limited, owing to the relatively low concentrations of molybdenum in the reactor. Because the model is much less computationally expensive than large depletion codes, it was possible to perform the large number of simulations needed to determine the effect of reactor design parameters on molybdenum extraction. It was found that the concentration at which Mo was present in the reactor had a profound effect on the isotopic composition and the fraction of the Mo-99 that could be successfully extracted. If molybdenum is allowed to buildup in the reactor, the majority of Mo-99 is lost by decay or activation, leaving only a small fraction of Mo-99 in the extracted molybdenum. Therefore, separation processes that require a high relative molybdenum concentration for extraction are not desirable. On the other hand, separation processes become more difficult at low concentrations. Thus, a trade-off exists between the difficulty of the separation and effectiveness of the extraction as measured by the recovery fraction and product quality. It was also found that increasing the power density enabled improved Mo-99 recovery for separation processes at increased concentrations. These results demonstrate the use of a model to explore the viability of Mo-99 extraction from an MSR, the influence of operating parameters on that extraction, and the conditions under which Mo-99 extraction can be optimized.

534133 Compatibility of Ni-Cr Alloys in Static and Flowing Commercial Molten Chloride Salt

Abstract:

The next generation of concentrating solar power (CSP) plants will require thermal storage to enable dispatchable renewable energy. One storage option to achieve >50% electrical efficiency at >700°C is chloride salts. To evaluate the compatibility of commercial Ni-Cr alloys in $MgCl_2$-KCl-NaCl salt from a commercial source, a laboratory purification process was used to minimize the O, S and Br in the starting salt. The first stage evaluated commercial Ni-based alloys such as 230 (Ni-22Cr-12W) and 600 (Ni-16Cr-9Fe) under isothermal conditions in a sealed Ni or Mo capsule at 600°-800°C. At 800°C, multiple experiments are being conducted with the same conditions to evaluate reproducibility of mass change as well as depth of Cr dissolution, which is the typical alloy degradation mechanism in halide salts. The second stage of the work will evaluate the compatibility of alloy 600 in flowing salt using a thermal convection loop where a temperature gradient is more representative of the CSP application. This research is sponsored by the U.S. Department of Energy, Office of Energy Efficiency and Renewable Energy, Solar Energy Technology Office.

530200 Investigating Chromium Dealloying in Molten Chloride Salt

Abstract:

It is well understood that selective dissolution of chromium is the primary mechanism by which structural alloys degrade in molten salt. The rate of dissolution depends on Cr solubility in the salt, and on the presence of impurities. A more complete knowledge of chromium dealloying is desired for the design of new materials, and for regulating salt conditions in molten salt reactors and concentrating solar power systems.

For this work, capsule-based corrosion experiments were conducted in which commercial alloys, Alloy-N variants, and a series of Ni-base alloys with Cr contents ranging from 7% to 24% were exposed to molten KCl-$MgCl_2$ or NaCl-$MgCl_2$ at 700°C for up to 2000 hours. Samples were weighed to determine mass change, mounted metallographically, and characterized with optical microscopy and SEM to characterize the exposed surfaces. SEM-EDS was used to measure the extent of Cr depletion near sample surfaces. The salts in which the samples were exposed were characterized with ICP-OES to understand the effect of salt Cr content on the corrosion rate of the binary alloys.

This talk will report on new insights gained into selective dissolution of Cr in molten chloride salts

This work was sponsored by the Laboratory Directed Research and Development Program of ORNL, managed by UT-Battelle, LLC, for the U.S. Department of Energy.

516965 Electrochemical Deposition of Sr and Ba into Liquid Bi from Molten Salt Electrolytes

Abstract:

Electrochemical deposition of Sr and Ba into liquid Bi metal was investigated in LiCl-KCl-$SrCl_2$-$BaCl_2$ electrolytes at 500 °C as a means to separate stable alkaline-earth ions from the eutectic LiCl-KCl utilized for recycling used nuclear fuel, by leveraging the strong chemical interactions between alkaline-earth metals and liquid Bi. The liquid Bi electrodes were subjected to cathodic discharge in eutectic LiCl-KCl with the addition of 5 mol% total of $SrCl_2$ and/or $BaCl_2$. The use of Bi resulted in co-deposition of Sr (2.0–6.5 mol%), Ba (4.1–12.8 mol%), and Li (5.9–16.2 mol%). The observed co-deposition was supported by thermodynamic analyses of electrode potentials by incorporating the experimentally determined activity values of each alkali/alkaline-earth metal in Bi. Furthermore, the contents of $SrCl_2$ or $BaCl_2$ in the electrolyte was varied from 5 to 0.1 mol% to observe the lower limit of alkaline-earth salt composition required to deposit Sr or Ba. The results of this work suggest that alkaline-earth fission products accumulated in molten salts (Sr^{2+} and Ba^{2+}) can be recovered into liquid Bi by electrochemical separation, which could be employed as a critical step for recycling the process salt (LiCl-KCl) in order to minimize the generation of additional nuclear wastes.

525498 Electrochemical Separation of Cs from Molten Salts Using Liquid Metal Electrodes

Abstract:

The feasibility of removing Cs from LiCl-KCl-CsCl (eutectic-3 mol%) using liquid Bi, Sn, and Pb electrodes was examined via thermodynamic properties measurements as well as electrochemical depositions. The accumulation of Cs^{137} in the LiCl-KCl-UCl_3 electrorefiner electrolyte leads to frequent replacement and disposal as high level nuclear waste due to the high heat density and short half life of Cs^{137}. Unfortunately, in the chloride system Li and K are less noble than Cs, making Cs difficult to remove from a LiCl-KCl-based electrolyte. However, by leveraging the strong atomic interactions between Cs and liquid metals such as Bi, Sn, and Pb, we examine the possibility of selectively removing Cs from chloride-based molten salt, providing a pathway to reducing this particular source of nuclear waste. Electrochemical depositions on these liquid metal working electrodes were performed up to a depth of 100 C g^{-1}, resulting in co-deposition of Cs and Li up to (4 mol% Cs - 12 mol% Li) at 500 °C in Bi. Electrochemical deposition experiments were complimented by electromotive force measurements for the Cs-Bi system to elucidate the thermodynamic properties (namely, the activity of Cs in Bi) leading to the observed shift in expected deposition behavior due to the liquid metal electrode. This work suggests the possibility of using liquid metal electrodes to remove Cs from molten salts and present a potential path towards reducing the volume of nuclear waste generated from the electrorefining process.

527292 Electrochemical Analysis of Bi-Analyte Electrorefiner Salt with High Concentration of UCl_3

Abstract:

During the spent nuclear fuel electrorefining process, the spent fuel is oxidized into the molten salt mixture followed by selective reduction on a cathode. Consequently, a large inventory of fissile actinides is held up in the salt. Thus, an important requirement for operating the process is to keep track of the actinides in the salt mixture to prevent proliferation. Several recent publications on either low UCl_3 concentration or single analytes show that electrochemical senor could potentially monitor salt concentration in real time. However, the UCl_3 concentration in the electrorefiner is often between 5-10 wt% during operation. Therefore, the objective of this research is to focus on bi-analyte systems with up to 10 wt% UCl_3. Two bi-analyte systems U/Mg and U/Gd were studied. Mg is a surrogate for Pu due to their similar reduction potentials, and Gd is selected to represent the rare earth metals in fission products. In this research, normal pulse voltammetry (NPV) was selected to prevent working electrode surface area growth caused by U deposition. Optimizations were made to the procedure to make NPV compatible and highly accurate for this application. Through the study of U/Mg system, NPV parameters were established, and it was learned that migration currents are comparable in magnitude to diffusional currents. Solution resistance compensation and small working electrode were applied for further optimization during the study of U/Gd system. The final concentration prediction average error values for UCl_3 and $GdCl_3$ are 1.6% and 2.70% respectively. This result shows that NPV is a suitable method for monitoring spent nuclear fuel electrorefiner molten salt metal concentrations in real time. It is also a promising approach for monitoring actinide concentrations in molten salt reactors.

528296 Monitoring of Actinide Concentrations in Molten LiCl-KCl Salt Using Alpha Spectroscopy

Abstract:

In order to develop a sensor for measuring actinide concentrations in molten electrorefiner salt, a process for electrodeposition of U and Th from LiCl-KCl-UCl_3-$ThCl_4$ onto Pt-coated surfaces has been developed. To facilitate the counting of alpha particles by the sensor, the objective is to electrodeposit a uniform, one micron thick layer of actinides. Such a deposit was accomplished using repeating chronoamperometry (RCA) tests in LiCl-KCl-UCl_3, LiCl-KCl-$ThCl_4$, and LiCl-KCl-UCl_3-$ThCl_4$ molten salt mixtures at 500°C. Uniformity of the deposits was verified using scanning electron microscopy with energy dispersive x-ray spectroscopy (SEM-EDXS). Time to pass a fixed amount of electric charge through the cell was correlated with concentrations of UCl_3 and $ThCl_4$ in the salt mixtures. As expected, the deposition time was inversely proportional to actinide concentration in the salt. The actinides could be stripped and re-deposited multiple times on a given Pt-coated coupon with repeatable time results. This method shows great promise for simultaneously measuring concentrations of multiple actinides with very close electrochemical reduction potentials. It thus offers potential functionality beyond what is possible with only electrochemical methods such as cyclic or normal pulse voltammetry.

532816 Thermodynamic Properties of Strontium-Lead Alloys Determined By Electromotive Force Measurements

Abstract:

During electrorefining of metallic used nuclear fuel, alkaline-earth fission products (Sr and Ba) are oxidized at the anode and accumulate in the molten salt electrolyte. To evaluate the viability of electrochemically separating these stable ions from the electrolyte using a liquid Pb cathode, the thermodynamic properties of Sr-Pb alloys were investigated by electromotive force (emf) measurements. A Sr(s)|CaF_2-SrF electrochemical cell using a solid state electrolyte (binary CaF_2-SrF
emf values at 773 K-1073 K for ten Sr-Pb alloys at mole fractions $0.07 \leq x_{Sr} \leq 0.45$. These emf measurements were used to determine thermodynamic properties, including activity, partial molar enthalpy, and partial molar entropy. At 873 K, activity values of Sr in Pb were as low as 1.69×10^{-9} at mole fraction $x_{Sr} = 0.07$, implying strong atomic interactions between Sr and Pb. Phase characterization utilizing X-ray diffraction (XRD) was conducted on alloy compositions $0.07 \leq x_{Sr} \leq 0.75$ and thermal analysis was conducted on compositions up to $x_{Sr} = 0.34$ using differential scanning calorimetry (DSC). Combining phase analysis and transition temperatures, the Sr-Pb phase diagram was refined. Based on the measured thermodynamic properties, the electrochemical deposition into a liquid Pb cathode was conducted in molten LiCl-KCl-$SrCl_2$ at 500°C and the Pb cathode was characterized to examine the deposition products (e.g., Sr, Li, and K).

523953 Evaluation of Concentrations of Residual Water, Hydroxides, and Oxides in Molten Anhydrous $CaCl_2$

Abstract:

Molten calcium chloride can be used as an electrolyte for reduction and electrorefining of actinides. However, its hygroscopic nature may present a barrier to this application. Simply heating the salt to its molten state may not entirely remove the water. At elevated temperature, the water may hydrolyze $CaCl_2$ to form calcium hydroxide or oxideâ€"both of which are expected to have substantial solubility in the molten salt. Such impurities are expected to interfere with the designed oxidation/reduction reactions in the electrorefiner. In this study, we examined a number of methods for quantifying these contaminantsâ€"including thermogravimetric analysis, cyclic voltammetry, and acid-base titration. The starting material was commercially prepared ultradry $CaCl_2$, which was melted to 850Â°C. Thermogravimetric analysis of this salt indicated a 1 wt% mass loss on heating to 770Â°C, indicating a water content of at least that level. Cyclic voltammetry of the as-received $CaCl_2$ indicated the presence of a reduction peak that could be attributed to calcium hydroxide. To verify this assumption, current peak response to known additions of calcium hydroxide was recorded. The resulting correlation can be used to make in-situ measurements of calcium hydroxide concentration in molten $CaCl_2$. Acid-base titration yields the total concentration of $Ca(OH)_2$ and CaO in solidified samples of the salt. The combination of these three methods can be used to determine removable residual water as well as hydroxides and oxides that form from non-removable water. This presents a route for future work in $CaCl_2$ molten salts towards developing an effective drying process.

538889 A Lightweight Betaemitter for Power Applications

Abstract:

Betavoltaic batteries have been around for quite some time but have yet to be manufactured with enough efficiency to safely produce an equivalent battery to current low power batteries. A betavoltaic battery produced with tritium has the potential to produce a battery with a life of around 12 years. Lithium can be easily intercalated into highly-oriented pyrolytic graphite (HOPG). The trapped lithium can then be neutron-irradiated to form tritium. This would produce a HOPG structure with tritium infused throughout. Understanding the lithium-intercalated HOPG structure is important to determining what might happen upon irradiation. Different methods of analysis were used for characterizing the lithium intercalated HOPG. Auger electron spectroscopy was used to determine if lithium was present. X-ray diffraction determined the spacing between the graphene planes. Although scanning electron microscopy proved inconclusive, transmission electron microscopy and diffraction provide valuable information about the lattice.

532527 Effective Removal of Pertechnetate from Groundwater By Bimetallic Porous Iron

Abstract:

^{99}Tc is a major long-lived fission product generated during nuclear power generation. Over the years, Tc has been inadvertently introduced into the environment from leaks at waste storage facilities and currently is one of the key risk drivers at the US DOE sites. The most common chemical form of Tc in liquid nuclear wastes or in the environment is pertechnetate (TcO_4^-). TcO_4^- displays limited adsorption to common sediment minerals and is highly mobile making it difficult to be immobilized. As the stockpile of ^{99}Tc-bearing nuclear waste continues to increase rapidly, novel sequestration technologies are needed to reduce its potential contamination of the environment and living organisms.

We developed highly efficient and cost-effective bimetallic porous iron materials that were investigated for TcO_4^- sequestration from groundwater. Catalytic bimetallic nanoparticles were grown onto porous iron by chemical reduction methods. The bimetallic porous iron materials were characterized by several analytical techniques to elucidate their morphology, size, composition, and porosity. Specifically, we used X-ray diffraction, scanning electron microscopy, energy dispersive X-ray spectroscopy, and BET surface area using Kr as the adsorption probe. The bimetallic porous iron materials were evaluated for removal of TcO_4^- and its surrogate (ReO_4^-) from artificial groundwater under ambient conditions using batch experiments. Tc and Re chemical speciation and molecular removal mechanisms were investigated using synchrotron radiation X-ray near-edge structure spectroscopy.

The results show that porous iron is >92% Î±-iron with small amounts of Fe and Al oxides. It has a sponge-like porous structure with a BET surface area of 0.95 m^2/g, pore volume of 0.0068 mL/g, and average pore diameter of 364 Ã…. When evaluated for contaminant removal, the porous iron successfully removed >97% present in an artificial groundwater solution containing 3.2×10^{-6} M TcO_4^-. In contrast, conventional granular zero valent iron, tested as a control, was entirely ineffective with 0% Tc removal. The removal capacity of the porous iron was demonstrated to be 27 mg Tc/g iron. Further, Tc K-edge X-ray absorption spectra of the porous iron samples retrieved from Tc batch experiments showed that both Tc(IV) and Tc(VII) species are present. Linear combination fitting indicated that 50-70% of the Tc was reduced to Tc(IV) that was incorporated into the secondary Fe oxides. In conclusion, these results indicate that porous irons are highly efficient materials for the immobilization of TcO_4^- present in groundwater through both adsorption and chemical reduction mechanisms. They might also be novel materials for effectively immobilizing a wide array of other radioactive contaminants in the environment.

530609 Migrating High Performance Computing to the Amazon Cloud - Methods and Examples Using Environmental Systems Analysis

Abstract:

To adequately analyze a subsurface environmental system, including assessing the inherent uncertainty in the information available to characterize it, and the ability to robustly include the governing physical processes requires numerical computational models. Since these models require inclusion of complex biogeochemical processes and simulate large aerial extents, each model run can require execution times on the order of hours or days to complete.

Adequate analysis of the uncertainty inherent in these subsurface systems performances in their natural state, or due to imposed stresses such as experienced during active dewatering or groundwater remediation, can require 100s or 10,000s of simulations be conducted. Thus, an analysis of complicated subsurface systems can require 100,000s of hours of computer processing time.

With the advent of Cloud computing, computation time on an as-needed basis is readily available. This work discusses implementing large-scale computational analysis on the Amazon Cloud including determining a capacity plan, configuring the project workload, managing the grid engine worker queue, staging and storing results for post processing. Three examples of various size and complexity of actual Cloud computation projects are provided with experience and lessons learned.

544065 Design Optimization for Environmental Projects - A Primer on Optimization Methods

Abstract:

Optimal project solution design for environmental projects involves selecting the appropriate values for any number of design parameters. This presentation provides the practitioner with a general set of tools that are available for use on a wide variety of project types.

536128 Quantifying Shifts in Trace Element Emissions from Coal-Fired Power Plants

Abstract:

Coal contains a variety of trace elements that are volatalized when the coal is combusted in coal-fired power plants (CFPPs). These trace elements are emitted to the environment in the exhaust gas, ashes, or in wastewater effluents. The partitioning of trace elements to the gas, solid, or liquid phase is determined to a large extent by installed air pollution control devices (APCDs). CFPPs have changed their installed APCDs and the coal they combust in response to changes in environmental regulations. We combine data from the U.S. Energy Information Administration and United States Geological Survey to build a model that will estimate the solid, liquid, and gas phase emsisions of seven trace elements (As, Cl, Se, Hg, Br, B, and Pb) from all US CFPPs with a capacity greater than 10 MW. We use this model to quantify how the installation of new APCDs has shifted trace element emisssions from the gas phase to the solid and liquid phase over past decades. We also calculate the impact of the shift towards low-sulfur coals on trace elements in the coal combusted at CFPPs and how alternative, hypothetical environmetnal compliance strategies would have produced different trace element emissions patterns.

527960 Adsorption of Cadmium, Nickel, and Lead on Modified Clinoptilolite:Â Equilibrium, Kinetic and Selectivity Studies

Abstract:

Environmentally friendly treatment methods for heavy metals prior to discharge are considered favourable so as to protect ecosystems. The performance of modified zeolite from two different sources (South African and the USA) for the adsorption of Ni^{2+}, Cd^{2+} and Pb^{2+} from synthetic industrial effluent contaminated with metal concentration levels at 50, 150 and 500mg/L was assessed. The selectivity of the zeolite for the adsorption of Ni^{2+}, Cd^{2+} and Pb^{2+} was investigated at mixed feed solutions containing all three ions in equal concentrations and single component concentrations containing only one of the ions. The homoionic forms of the zeolite were made Na^{+}, K^{+} and Ca^{2+}. Batch experiments were then conducted to measure the uptake of metals by the zeolites. The zeolites were characterised using SEM, XRD and BET. The South Africa clinoptilolite has a higher surface area and pore volume ($17.52m^2/g$ and $0.047cm^3/g$ respectively) than the USA zeolite ($12.26m^2/g$ and $0.028cm^3/g$ respectively) for the Na^{+} homoionic form. According to the equilibrium studies, the selectivity sequence can be given as Pb^{2+} >Cd^{2} >Ni^{2+}, with good fits being obtained using Langmuir and Freundlich adsorption isotherms for low metal concentrations as compared to higher concentrations. The adsorption was found to depend on the homoionic form of the zeolite and to a lesser extent the source of the zeolite. The selectivity of a particular zeolite for a particular heavy metal can be altered by the homoionic form of the zeolite. Generally, the adsorption capacity of the USA clinoptilolite is higher than the capacity of the SA clinoptilolite revealing the potential of clinoptilolite in industrial effluent treatment.

539974 Direct Feed Low Activity Waste Test Platform â€“ Alternative Demonstrations

Abstract:

The Low-Activity Waste Pretreatment System (LAWPS) facility is planned to pretreat Hanford tank waste supernate by filtering solids and processing through ion exchange columns to remove cesium. The filtration process is achieved through dead end filtration. The effluent from ion exchange will be vitrified to produce LAW glass.

The goal of this effort was to demonstrate alternative processing approaches for filtration, ion exchange and melter off-gas treatment. To that end, dead end filtration and alternative ion exchange media were tested. Actual waste samples were retrieved from Hanford Site Tank AP-107 and were composited for testing. These composited samples were first tested with cross flow filtration and dead end filtration to determine the filter flux under typical LAWPS conditions. The material was subsequently treated for cesium removal with CST followed by vitrification using a flow through melter system. These results of these tests will be described.

516311 Improved Foam Control during High Level Radioactive Waste Processing

Abstract:

Antifoam is needed during processing of HLW due to the high gas generation from steam and chemical reaction offgas products to prevent foam from contaminating the condensate. Antifoam is used to minimize foam[1] during chemical processing and HLW evaporation at SRS[2] and Hanford[3]. The antifoam used in the Defense Waste Processing Facility (DWPF) impacts flammability during chemical processing (generates three flammable degradation products) and while feeding the melter[4-5] (can decompose to CO/hydrogen). It also is the likely source of methyl for the methyl mercury present in the tank farm and excessive mercury in Saltstone[6]. The planned startup of The Salt Waste Processing Facility (SWPF), with much higher throughput, will challenge DWPF to process more efficiently.

DWPF employs Antifoam 747[7], a superspreader produced by Momentive Performance Materials, as an antifoaming agent during waste processing. Despite its ability to control foam, processing issues have risen from use, which has led to two â€œPotential Inadequacy in the Safety Analysesâ€ (PISAs) and has limited DWPF facility throughput.

During DWPF chemical processing, antifoam must be effective up to 103ËšC between pH of 3-13. Antifoam 747 is most effective at pH 6-8[8] and degrades as pH deviates. SRNL identified three flammable antifoam degradation[9] products using mass spec and FTIR offgas analyzers during experiments[10]. Equipment has not been used in identifying decomposition products from antifoam agents used elsewhere.

A new Antifoam or a new method to control foam is needed to minimize DWPF processing time. The progress in developing the new foam control method will be discussed.

1. Bindal, S. K.; Nikolov, A. D.; Wasan, D. T.; Lambert, D. P.; Koopman, D. C., Foaming in Simulated Radioactive Waste. *Environ. Sci. Technol.* **2001,** *35* (19), 3941-3947.
2. Wasan, D. T.; Lambert, D. P. *Foaming and Antifoaming in Radioactive Waste Pretreatment and Immobilization*; Illinois Institute of Technology: Chicago, IL, 2001.
3. Calloway, J. T. B.; Martino, C. J.; Jantzen, C. M.; Wilmarth, W. R.; Stone, M. E.; Pierce, R. A.; Josephs, J. E.; Barnes, C. D.; Daniel, W. E.; Eibling, R. E.; Choi, A. S.; White, T. L.; Crowley, D. A.; Baich, M. A.; Johnson, J. D.; Vijayaraghavan, K.; Nikolov, A. P.; Wasan, D. T., Radioactive Waste Evaporation: Current Methodologies Employed for the Development, Design and Operation of Waste Evaporators at the Savannah River Site and Hanford Waste Treatment Plant. **2003,** (37327), 157-170.
4. Choi, A. S.; Smith III, F. G.; McCabe, D. J. *Preliminary Analysis of Species Partitioning in the DWPF Melter. Sludge batch 7A*; SRNL-STI--2016-00540; Savannah River National Laboratory: Aiken, SC, 2017.
5. *Evaluation of the Safety of the Situation (ESS): Melter Feed Rate Temperature Correlation Basis (PISA PI-2014-0009)*; Savannah River Remediation LLC: Aiken, SC, 2016.
6. Meraw, H. J. *Volatilization and Flammability Characteristics of Elemental and Organic Mercury*; X-ESR-G-00048, Revision 2; Savannah River Remediation LLC: Aiken, SC, June 2015, 2015.
7. Koopman, D. C. *Comparison of Dow Corning 544 Antifoam to IIT747 Antifoam in the 1/240 SRAT*; WSRC-TR-99-00377; Savannah River Technology Center: Aiken, SC, 2000.
8. Lambert, D. P.; Koopman, D. C.; Newell, J. D.; Wasan, D. T.; Nikolov, A. P.; Weinheimer, E. K., Improved Antifoam Agent Study End of Year Report, EM Project 3.2.3. 2011.
9. McCord, J. B. *Evaluation of the Safety of the Situation (ESS): Melter Feed Rate Temperature Correlation Basis (PISA PI-2014-0009)*; Savannah River Remediation LLC: Aiken, SC, 2016.
10. Lambert, D. P.; Zamecnik, J. R.; Newell, J. D.; Martino, C. J. *Impact of Scaling on the Glycolic-Nitric Acid Flowsheet*; Savannah River National Laboratory: Aiken, SC, 2016.

536472 The Nature and Quantity of Solids Suspended in Hanford Nuclear Waste Supernatants

Abstract:

The Hanford Site contractors are currently designing and constructing a waste treatment process to treat highly concentrated sodium-bearing supernatants. The treatment process is expected to require suspended solids removal. In order to design the solids removal process, basic properties of suspended solids (size, density, chemical composition, shape, abrasively, quantity, etc.) are required. In order to provide insight in the solids characterization, a three pronged approach was taken: 1) Evaluate historic records on the solids concentration in supernatant grab samples; 2) characterize two samples that contained a floating solids layer at the surface; and 3) perform filtration studies of waste supernatants to characterize those solids that are not visible. The historic records indicated that the overwhelming majority of grab samples had no visible solids, however, 5 out of 163 had between 0.1 and 0.5 volume % solids. The characterization of floating solids indicated that they could be large and be smooth-surfaced aggregates. Particulates made of the aggregates could be up to 200 microns in diameter. Such solids were suspended by complex reactions of saponification, adsorption of organics, and available surface area for the adsorption of organics. The filtration studies determined that other suspended solids were simply very small particles, usually less than 2 microns in diameter, that were extremely slow to settle. These include both organic and metalloid particles.

526720 Dechloriation of Electrorefiner Chloride Salt Via Ion-Exchange Using Ultra-Stable H-Y Zeolite

Abstract:

A dechlorination process has been designed to generate durable waste forms from spent nuclear fuel electrorefiner salt waste. It is based on ion exchange of the radioactive chloride salt with dehydrated protonated ultrastable Y-type (USHY) zeolite to form metal exchanged zeolite and gaseous HCl. This process has been demonstrated using both LiCl-KCl and surrogate electrorefiner salt in a small-scale rotary tube furnace. Dechlorination was confirmed and measured using both free chloride ion analysis of samples of the ion exchanged zeolite and output of an autotitrator that continuously balances HCl generated in real time. Preliminary results showed that over 90% dechlorination was achieved in 48 hr using pelletized USHY (Si/Al=2.6) at 625^{0}C in a continuous rotary tube furnace without zeolite structure collapse. Temperatures of 650°C or higher result in rapid collapse of the zeolite structure into an amorphous phase. The rate of the ion exchange is invariant when reducing particle size below 250 micron diameter, indicating that particle mixing is likely the rate limiting process. In addition to electrorefining waste processing, this technology could be used to generate waste forms from salts used in a chloride molten salt reactor (Cl-MSR).

531021 Iodine Adsorption in Reduced Agz and Ag-Aerogel in Presence of Water and NO_X

Abstract:

Reprocessing used nuclear fuel (UNF) generates several volatile radioactive species, including ^3H, ^{14}C, ^{85}Kr, and ^{129}I, along with significant amount of water vapor and NO_X into the off-gas streams. The presence of water vapor and NOx impacts the performance of the adsorbents for capturing the radioactive gases, especially, the Ag-containing adsorbents for iodine adsorption. In this study, the iodine adsorption performance of reduced Ag mordenite (Ag^0Z) and Ag functionalized silica aerogel (Ag^0-aerogel) in the presence of water and NO_X were investigated experimentally and analytically. Through thin-bed adsorption experiments, uptake curves of I_2 on Ag^0Z and Ag^0-aerogel were obtained in gas streams containing I_2, H_2O, and NOx. Through chemical analyses of the iodine-loaded adsorptions including scanning electron microscopy (SEM), X-ray powder diffraction (XRD), X-ray photoelectron spectroscopy (XPS), Raman spectroscopy, and X-ray absorption fine structure (XAFS), the binding information of Ag with I_2, H_2O and NO_X during the adsorption process were obtained. Several chemical reactions occurred during the process forming silver iodide, silver nitrates and silver sulfate (in Ag^0-aerogel). Their chemical compositions were also determined. It was found that the oxidation of Ag by H_2O and NO_x occurs much faster than the iodine adsorption process due to the much higher concentration of H_2O and NO_X. As a result, the iodine adsorption performance of Ag^0Z and Ag^0-aerogel decreased significantly.

531056 Kinetic Study of Ag Mordenite and Ag Functionalized Silica Aerogel Aging in Nuclear Fuel Reprocessing Off-Gases

Abstract:

In the study of spent nuclear fuel reprocessing off-gas treatment, radioactive iodine vapor can be effectively captured by silver containing adsorbents. However, the impact of aging on adsorbents performance is one major issue during off-gas treatment operation due the presence of water vapor and NOx. To better understand the performance of the adsorbent in the off-gas treatment system, it is important to study the aging mechanism and kinetics. In this study, the impact of aging from off gas components on the kinetics of iodine capture on silver exchanged mordenite (Ag^0Z) and silver functionalized silica aerogel (Ag^0-aerogel) were investigated, two well recognized adsorbents for capture of radioactive iodine. These adsorbents were aged in dry air, humid air, 1% NO/N_2, and 2% NO_2/air, at 100 °C, 150 °C, and 200°C. Consequently, iodine loading experiments with the aged adsorbents were conducted. Results showed that humid-air, 1% NO/N_2, and 2% NO_2/air had strong impacts on the performance of Ag^0Z and Ag^0-aerogel. It was also found that the negative aging effect increased as the aging temperature increased. For example, the Ag^0Z aged in humid air showed 42% , 45% and 58% of iodine loading capacity losses at 100°C, 150°C, and 200°C, respectively. Kinetic models were developed and evaluated to simulate the aging processes in different gases. The models include reversible and irreversible n^{th} order reaction models to describe diffusion and reaction of the aging gas into the adsorbent pellets. Related parameters such as reaction rate constants and diffusivities were obtained. Work is in progress to incorporate the aging kinetic models into the iodine adsorption models for fixed beds to predict the actual adsorption process in an off-gas stream.

538319 Microflow Visualization of Tri-n-Butyl-Phosphate/Dodecane and Nitric Acid in a Centrifugal Contactor

Abstract:

Despite the common use of centrifugal contactors in nuclear solvent extraction applications, much is unknown about the underlying microflow structure. This lack of knowledge hampers the development of a modern continuum theory able to predict mixing and transport phenomena for solvent extraction processes. The ability to observe the micromixing flow and distinguish the motion and configuration of the phases involved is of great importance theoretically and practically. To that end, we report on direct, high-speed visualization of the microflow structure for the tri-n-butyl-phosphate (TBP)/dodecane/nitric acid system. Further insight is obtained by simultaneous measurement of the electrical resistance of the phase mixture to support visual observations.
This work sheds light on a fluid microflow region at the foot of the mixing zone of the contactor resolving the microsecond/micrometer motion of the continuous and dispersed phases.
The system in this study is the most prevalent solvent extraction combination used in the nuclear reprocessing industry to extract metal ions from spent nuclear fuel from a nitric acid aqueous solution into an organic mixture of TBP and dodecane. The purpose of this work is to describe salient features of the behavior of this system, heretofore unknown, to guide the development of a predictive modeling and simulation approach.

This work also offers unique guidance and insights for continuum-based modeling of the multiphase mixture mechanics. In particular what aspects of a mixing theory may provide qualitative and/or quantitative predictions of the rich phenomena reported here. No multiphase continuum mixture theory is generic enough for all flows;
therefore, this work is fundamental in guiding the future development of a predictive theory, modeling and simulation effort at the continuum level. In fact, the original motivation for this work was to provide experimental information for building an appropriate mixing theory for the system at hand. Finally, advances in rigorous modeling of this system will likely provide the basis for other similar multiphase, multicomponent fluid transport used in nuclear recycling processes.

AUTHOR INDEX